Thermodynamics and Statistical Mechanics

Thermodynamics and Statistical Mechanics

G. SOCRATES, M.Sc., Ph.D., M.Inst.P.
Brunel University, London

LONDON
BUTTERWORTHS

THE BUTTERWORTH GROUP

ENGLAND
Butterworth & Co (Publishers) Ltd
London: 88 Kingsway WC2B 6AB

AUSTRALIA
Butterworth & Co (Australia) Ltd
Sydney: Pacific Highway Chatswood, NSW 2067
Melbourne: 343 Little Collins Street, 3000
Brisbane: 240 Queen Street, 4000

CANADA
Butterworth & Co (Canada) Ltd
Toronto: 14 Curity Avenue, 374

NEW ZEALAND
Butterworth & Co (New Zealand) Ltd
Wellington: 26-28 Waring Taylor Street, 1
Auckland: 35 High Street, 1

SOUTH AFRICA
Butterworth & Co (South Africa) (Pty) Ltd
Durban: 152-154 Gale Street

First published 1971
© Butterworth & Co (Publishers) Ltd, 1971

ISBN 0 408 70179 X Standard
0 408 70193 5 Limp

Filmset by Photoprint Plates Ltd,
Rayleigh, Essex

**Printed by photo–lithography and made in Great Britain
at the Pitman Press, Bath**

Preface

The object of this text is to provide the reader with a comprehensive introduction to the basic concepts of thermodynamics and statistical mechanics. It is intended for science and engineering undergraduates.

The emphasis throughout is on simplicity and clarity, the order of presentation and the explanations given being those which have been found most successful with students and acceptable to them. It is often found that ideas are presented to students which are not wholly true. This may either be in order to gloss over certain difficulties or possibly from a lack of understanding. Despite the simplicity of approach of the present text, the truth has not been overlooked.

SI units have been used, and for the convenience of those unfamiliar with them an appendix on this topic has been included.

The approach to thermodynamics is, of necessity, mathematical. Quite often, students lose themselves in the mathematics, not appreciating that it is merely a tool used to obtain the end-result. Basically, all the mathematics that is required is a little algebraic manipulation and a small amount of elementary calculus.

The classical approach to entropy of Chapter 4 may be omitted with no loss of continuity, this being provided by the alternative approach of Chapter 5. The modern practice of chemists is to avoid the classical approach. Probability has been introduced in Chapter 5 in order to clarify the concept of entropy. The third law is presented soon after the first and second laws, rather than being left as an afterthought, as is often the case. All types of solutions – gas, liquid and solid, ideal and non-ideal – are dealt with as a whole in Chapter 13, thus avoiding unnecessary repetition. Because of the importance of chemical equilibria, an elementary derivation of the van't Hoff isotherm is given in Chapter 8 in order that the reader be familiar with it at an early stage. In addition, a more general approach is given in Chapter 14.

Asterisks have been employed at section headings to indicate sections of interest to readers wishing to have a more thorough knowledge. These may be omitted without any loss of continuity.

I should like to acknowledge the assistance given to me by many colleagues and past students, and to thank Dr. K. P. Kyriakou for his encouragement. It is also a great pleasure to acknowledge the vast amount of assistance I have received from my wife, Jeanne, not only in the form of helpful discussions and encouragement, but also in the typing of the manuscript.

Brunel University   G.S.
April 1971

Contents

1
Introductory Concepts

1.1 Introduction

In thermodynamics, when we talk of a *system* we mean a specific portion of matter, with definite boundaries, on which our attention is focused. Other terms which will be referred to frequently are:

Surroundings. All space and matter external to the system, which may or may not interact with the system.

— *Open systems.* Those systems in which mass may be transferred between the system and the surroundings.

— *Closed systems.* Those systems of constant mass, i.e. in which no transfer of mass occurs.

Universe. The system plus its surroundings (no celestial connotations being implied).

Isolated systems. Systems which cannot exchange energy of any kind, including mass, with their surroundings. In other words, an isolated system is not allowed to exchange heat, work or mass with its surroundings.

A process. Any transformation of a system from one equilibrium state to another over an interval of time. (Work and heat transfers may possibly occur during a process.)

The path of a process. The specific series of equilibrium states through which the system passes during the process.

A complete description of a system defines the existing condition or *state* of the system. There are two ways of giving a complete description of a system, microscopic and macroscopic. For simplicity, let us consider a system that consists of a single homogeneous substance. (By homogeneous, we mean chemically and physically uniform throughout.)

1

MICROSCOPIC DESCRIPTION

A homogeneous system may be thought of as consisting of a very large number of particles (atoms or molecules) of the same mass. The position and velocity of each particle are both needed for a complete description. That is, for each particle we need three cartesian coordinates x, y, z and the velocity components v_x, v_y, v_z. Therefore, for N particles we need $6N$ values to determine the state of the system. The state defined in this manner is known as a microstate. Because of the motions of the molecules and atoms this type of description would only be true for a single instant in time. The microscope description leads to a *statistical* approach to thermodynamics.

MACROSCOPIC DESCRIPTION

To give this type of description, we need a knowledge of the properties of the system such as mass M, pressure P, volume V, temperature T, etc. These measurable properties may, in theory, be determined from a knowledge of the microscopic state. For instance, pressure is the average force due to the impact of molecules on unit area, i.e. the average rate of change of momentum for all the molecular impacts on unit area. A state defined by using measurable quantities is known as a macrostate.

The *microscopic description* has the following *disadvantages*:

(1) We assume a knowledge of the nature of matter, for example that the system is composed of molecules;
(2) The description of the state of the system requires the knowledge of a large number of values ($6N$);
(3) The quantities required to describe the system (e.g. the positions and velocities of all the molecules) cannot be easily determined;
(4) The description is only true for a single instant in time.

An *advantage* of the microscopic approach, and therefore a disadvantage of the macroscopic approach, is that the conceptual insight on the molecular scale given by statistical thermodynamics cannot be attained in any other way.

The *advantages* of the macroscopic description are:

(1) No assumption about the composition of matter is necessary;
(2) Only a relatively small number of quantities is required to describe the state of the system;
(3) The description is given in terms of quantities which can be easily measured (M, P, V, etc.).

Of course, the results obtained from both the microscopic and macroscopic approaches to thermodynamics must agree.

1.2 Functions of state

Consider a system consisting of a single homogeneous substance. In giving a precise macroscopic description of the state of the system, we could use the properties mass M, volume V, pressure P, temperature T, density ρ, viscosity η, refractive index n, etc. These *functions of state*, as they are known, are used to define the state of the system. Functions of state are independent of the previous history of the system. We say that a system is in an *equilibrium state* if its functions of state (i.e. properties) remain constant, provided that its external environment is not altered.

1.3 Extensive and intensive quantities

The properties of a system may be classed as intensive or extensive. If we consider a system which has been divided into two identical parts, then those properties which have not been altered by the division are *intensive* properties (e.g. pressure, temperature, density, viscosity, refractive index) whereas those properties which have been exactly halved are *extensive* properties (e.g. mass, volume, energy, charge). Intensive properties, as opposed to extensive properties, are therefore independent of the amount of material present. Extensive thermodynamic properties which we shall meet with in the course of this book are internal energy U, enthalpy H, entropy S, the Helmholtz function A, and the Gibbs function G.

Intensive analogues of extensive properties do exist, for example, specific volume v (i.e. volume per unit mass) or molar volume v (volume per unit amount of substance), molar internal energy u, molar enthalpy h, etc. Note that we use lower-case letters for the specific or molar quantities of functions, e.g. volume V and specific volume v, enthalpy H and specific enthalpy or molar enthalpy h, entropy S and molar entropy s. Specific or molar properties are all intensive functions of state.

1.4 Independent variables

The question now arises as to how many functions of state are needed to define the state of a system consisting of a single, homogeneous substance. From our own experience it is obvious that we do not need to know all the properties of the system. It has been found by experiment that, in general, if we fix the mass of the substance and any two other functions of state such as volume and pressure, or volume and temperature, or pressure and density, or volume and viscosity, then all the other functions of state are fixed automatically. That is, we need only know the values of three variables to define the state of the system. We therefore say that the system has *three independent variables*. By independent variable, we mean that the property cannot be calculated from a knowledge of the other functions of state used to define the state of the system. For instance, M cannot be calculated from P and T. On the other hand, density is not independent of M and V, and T^2 is not independent of T. At least one of the three independent variables must be an extensive property if the amount of substance present is to be known. For a system of *given* mass, only *two* independent variables are required to define the state. This means that if we choose any two independent variables (say P and V) then there exists an equation relating these variables and any one of the other functions of state (say T). That is, for a given mass, T may be considered a mathematical function of P and V only,

$$T = f(P, V) \tag{1.1}$$

Similarly, we may have

$$\eta = f(P, V)$$

and

$$\rho = f(P, V)$$

For a given system, these relations may be very complex, so much so that we may never know them explicitly.

Any function of state ψ may be written as a function of the independent variables of a system,

$$\psi = f \text{ (independent variables)} \tag{1.2}$$

Such equations are known as equations of state. We are, of course, all familiar with the equation of state of an ideal gas, i.e. for one mole,

$$Pv = RT$$

where T is the temperature on the *ideal gas temperature scale* (absolute zero on this scale being the temperature at which the volume of an ideal gas vanishes) and R is the *universal gas constant*.

Although, in general, only two functions of state are required to define the state of a pure homogeneous substance of given mass, there are systems for which *any* two properties will *not* define the state. For example, pressure and density are not sufficient to define the state of liquid water near 277 K, since the density of water goes through a maximum at this temperature.

1.5 Phase

A *phase* is a homogeneous part of a system separated from other parts by physical boundaries, i.e. a phase is physically and chemically uniform throughout. By a *heterogeneous system* we mean that the system is composed of more than one phase.

Example 1   In the system shown there are three phases (solid, liquid and vapour).

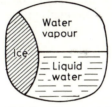

A solution (whether solid, liquid or gas) is considered as a single phase since it is chemically and physically uniform throughout.

Example 2   The system shown contains two phases.

Sometimes people use such terms as 'the gaseous state' and 'the solid state'. Strictly speaking, these are incorrect; one should refer to 'the gaseous *phase*' and 'the solid *phase*'.

1.6 Thermodynamic state

To define the thermodynamic state of a homogeneous system containing more than one substance, we must specify the amount of each component in the system and two other independent variables such as P and T.

To define the state of a heterogeneous system containing more than one substance, we must consider and describe each of the phases of the system. For each phase we must specify the content, i.e. the amount of each substance present, and two other independent variables.

1.7   Thermal equilibrium

A system is in *thermodynamic* equilibrium if the properties of the system do not alter so long as the external environment remains unchanged.

Consider two systems A and B, each in a state of equilibrium. Let them be brought into thermal contact by means of a *diathermic wall*, i.e. a wall allowing the transfer of heat from one system to another. The systems are thermally insulated from the rest of the universe by a surrounding adiabatic wall, i.e. a thermally insulating wall (see Figure 1.1). If the systems A and B were not in mutual equilibrium originally, then their properties would change until eventually

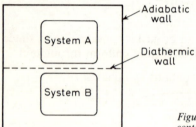

Figure 1.1 Systems A *and* B *in thermal contact and surrounded by an insulating wall*

they were so. Once this equilibrium is attained, no further change in the properties of the system will occur so long as the external conditions are not altered. The two systems are then said to be in *thermal* equilibrium.

It is possible to define a 'new' property which will show that the two systems are in thermal equilibrium, this property being known as temperature. By definition, two systems have equal temperatures if, when brought into thermal contact, their properties do not change. In other words, two systems in thermal equilibrium have the same temperature.

1.8   Zeroth law

Let us apply the concept of temperature equality of systems in thermal equilibrium to the following experiment. Two systems A and B, separated by an adiabatic wall, i.e. a thermally insulating wall, are both brought into thermal contact with a third system C by means of a diathermic wall (Figure 1.2).

Eventually, systems A and B will come into thermal equilibrium with system C and then there will be no further change in their properties. At equilibrium, systems A and C must have the same temperature, and similarly systems B and C must have the same

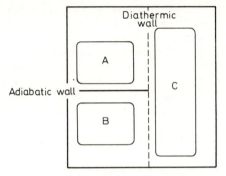

Figure 1.2. Systems A *and* B *in thermal contact with system* C *but insulated from each other*

temperature. Therefore, the temperatures of systems A and B must also be equal.

This experiment demonstrates the *zeroth law* of thermodynamics which states that *if each of two systems is in thermal equilibrium with a third, then they are also in thermal equilibrium with each other.*

Although this statement seems obvious, it cannot be derived from other observations of nature and is for this reason known as a *law.* The zeroth law is, of course, the basis of temperature measurement.

1.9 Temperature scales

Any system in a well-defined state may be used as a reference for temperature measurement, all other systems in thermal equilibrium with it having the same temperature. If a system is to be used as a thermometer, numerical values must be given to the temperatures of two well-defined states of the system and the thermometric property of the system must be measured in those states. A few of the commonest thermometric properties are volume, pressure, electrical resistance and length.

The Kelvin (or absolute) temperature scale, which is also known as the thermodynamic temperature scale, will be discussed later.

It is unique in that it is independent of the thermometric properties of any system and it will be shown to be identical with the ideal-gas temperature scale.

1.10   Thermodynamic equilibrium

For simplicity, consider a system that consists of a single homogeneous substance. If the properties of the system are uniform throughout, then, so long as the external conditions are unaltered, the system is said to be in *thermodynamic equilibrium*. For example, if the pressure were not uniform, turbulence would occur until the system reached equilibrium. It is only under equilibrium conditions that the state of a homogeneous system of given mass may be described by only two functions of state, each function having only one value.

A system is in thermodynamic equilibrium if it is in mechanical, thermal and chemical equilibrium.

Mechanical equilibrium exists if there are no net forces between the system and the surroundings and if also there are none within the system.

Thermal equilibrium exists if the temperature of the system is uniform throughout and equal to that of the surroundings with which it is in thermal contact.

Chemical equilibrium exists if the composition of each compound present is constant and if diffusion and solution do not occur.

1.11   Reversible and irreversible processes

A *quasi-static process* is one performed infinitely slowly.

A *reversible process* is a quasi-static process in which all means of energy dissipation are absent.

Energy dissipation may be due to friction, viscosity, electrical resistance, magnetic hysteresis, plastic deformation, etc. Reversible processes are hypothetical processes which are useful for comparison purposes. The concept of a reversible process may be compared with that of frictionless pulleys, weightless strings, etc. that are used in mechanics. Since the state of the system changes infinitely slowly during a reversible process, the system goes through a *continuous series of equilibrium states. A reversible process may be defined as one for which the system is always in equilibrium.*

Any process that is not reversible is known as an *irreversible process*. All naturally occurring (i.e. spontaneous) processes are non-equilibrium processes and therefore irreversible. We shall, in fact, show in Chapter 3 that irreversible processes cannot be reversed without leaving the surroundings permanently changed.

For the reversible expansion of a gas in a vessel fitted with a frictionless piston, the gas must always be in equilibrium, and therefore the pressure on the piston must be released infinitely slowly, so that at any instant in time the pressure and temperature are uniform throughout the gas. The state of the gas at any time may then be given by the single values of two variables, e.g. pressure 10^5 N m^{-2}, temperature 300 K.

If the piston were withdrawn rapidly, the gas would perform no work in rushing in to fill the vacuum left by the piston, pressure differences would be set up and turbulence would occur. In this case, the state of the gas cannot be given by the single values of two variables. Instead, an infinite number of values of the two variables would be required since the state of each point within the gas must be described. Only the initial and final states of the gas would then be equilibrium states. The rapid expansion described is an irreversible process.

The path of a reversible process can be shown by a line on a graph with suitable axes. However, we cannot do the same for an irreversible process since the state of the system cannot be described by the single values of two variables.

If the piston is released rapidly, (1) pressure differences will occur in the gas, and (2) the pressure opposing the change will increase. Both (1) and (2) will oppose the expansion and, hence, work is necessary to overcome them. For a *reversible expansion, the maximum possible work* is obtained since neither (1), (2) nor friction occur. Similarly, for a rapid compression, (1) and (2) would occur. Hence, for a *reversible compression, the minimum possible work* is used.

It is only for a *reversible* process that the *maximum work output* or the *minimum work input* can be achieved, that is, when the forces responsible for the change of state of the system are essentially always at equilibrium and energy dissipation does not occur.

In a reversible process, a system must be in thermal equilibrium. Therefore, if it is in thermal contact with its surroundings, it must be in thermal equilibrium with them at all times.

Since a system must be in equilibrium during a reversible process, all chemical reactions carried out under normal conditions are irreversible. Chemical reactions may be carried out reversibly only if the reaction is essentially at equilibrium. One very good and simple way of doing this is to construct an electrochemical cell in which the reaction that occurs produces an electromotive force (e.m.f.). If an opposing voltage is applied to the cell which is infinitesimally less than the e.m.f. of the cell, the reaction will proceed infinitely slowly. If the applied voltage equals the e.m.f. of the cell, no reaction will occur; if it is infinitesimally greater than the e.m.f., the reverse reaction will occur infinitely slowly.

Consider a system consisting of ice and liquid water in equilibrium at 273·15 K and 101 325 N m^{-2}. The ice may be melted reversibly by absorbing heat from a heat reservoir, the temperature of which is infinitesimally greater than 273·15 K.

A spring whose tension is increased infinitesimally will extend by an infinitesimal amount. If this tension is relaxed, the spring will return to its original position provided the elastic limit has not been exceeded. The work done in loading the spring equals the work recovered on unloading. The original extension was therefore reversible. If the elastic limit were exceeded, plastic deformation would occur and the spring would not return to its original position. Also, the work recovered would be less than that put in, the plastically deformed state being one of higher energy.

Some other examples of irreversible processes are:

(1) a metastable phase changing to the stable form of a substance (e.g. freezing of a super-cooled liquid, condensation of a super-saturated vapour);

(2) mixing of substances (e.g. mixing of gases or miscible liquids);

(3) transportation across a phase boundary (e.g. osmosis, dissolution of a solid);

(4) all naturally occurring processes, i.e. all spontaneous processes (e.g. all reactions occurring under normal conditions, the mixing of gases).

1.12   Work done by volume change

Consider a fluid in a cylinder with a frictionless piston. The fluid may do work on the surroundings by expansion.

Suppose the fluid at a pressure P expands by an infinitesimal amount dV, the piston moving by dx, against an opposing pressure P_{opp}. The work done, đw (the significance of the line through the work differential will be explained later), is given by

work done = force opposing the change × distance moved

$$đw = F_{opp} \times dx$$

We may write

$$đw = \frac{F_{opp}}{A} \cdot A \, dx$$

where A is the cross-sectional area of the piston.

Since the opposing pressure P_{opp} is the opposing force per unit area and $dV = A\,dx$,

$$\text{d}w = P_{opp} \cdot dV \tag{1.3}$$

The work done, w, in going from state 1 to state 2 is

$$w = \int_1^2 P_{opp} \cdot dV \tag{1.4}$$

By state 1 we mean that the pressure is P_1, the volume V_1, the

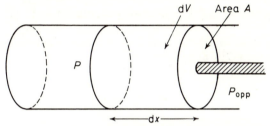

Figure 1.3. Fluid in a cylinder at pressure P is allowed to expand against opposing pressure P_{opp}

temperature T_1, etc., and similarly for state 2 we have P_2, V_2, T_2, etc.

If the *opposing pressure* is kept *constant* during the expansion, then

$$w = P_{opp} \int_{V_1}^{V_2} dV$$

$$w = P_{opp}(V_2 - V_1) \tag{1.5}$$

CONVENTION

Work done *by* a system is considered *positive*.
Work done *on* a system is considered *negative*.

For expansion, P_{opp} must be less than P. However, if P_{opp} is always only infinitesimally smaller than P, then the expansion will take place via a series of equilibrium states and, in the absence of friction, reversibly. For reversible expansion, we may say that P_{opp} is effectively equal to P ($P_{opp} = P$). Since, for expansion, P_{opp} must be less than or equal to P ($P_{opp} \leqslant P$), from Eq. (1.3) the work is a maximum during a *reversible* expansion and is equal to $P\,dV$. Similarly, for compression P_{opp} must be greater than P, and for reversible compression only infinitesimally greater than P (i.e.

$P_{opp} = P$). Therefore, from Eq. (1.3), the minimum work is done in a reversible compression.

The work done $đw$ by any infinitesimal expansion must be less than or equal to PdV, depending on whether the work is irreversible or reversible (indicated by I and R, respectively).

$$\overset{I}{\underset{R}{đw \leqslant PdV}} \tag{1.6}$$

For a finite process from state 1 to state 2,

$$\text{maximum work done during expansion} = \int_1^2 PdV \tag{1.7}$$

The work done in any expansion must be less than or equal to the work that would be done reversibly,

$$\overset{I}{\underset{R}{w \leqslant}} \int_1^2 PdV \tag{1.8}$$

Remembering the sign convention and multiplying this equation by -1, it is seen that the minimum work is done on a system in a reversible compression,

$$-w \geqslant -\int_1^2 PdV$$

Equation (1.4) gives the work done by an irreversible process in the absence of friction. In general, of course, P_{opp} may be time (and position) dependent. To illustrate that Eq. (1.4) holds for irreversible processes, consider the free expansion of a gas. In Fig. 1.4, one compartment of the vessel contains the gas, the other is evacuated. When

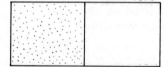

Figure 1.4.

the partition separating the compartments is removed, the gas expands without an opposing force, i.e. $P_{opp} = 0$. Therefore, no work is done in the expansion. This process of free expansion is irreversible since the pressure of any intermediate state is not uniform throughout the gas. Only the initial and final states are equilibrium states.

The amount of work that may be obtained from a gas must lie between the limits $\int_1^2 P dV$ and 0.

The unit of work and heat is the *joule*, which is defined as the work done in moving a mass of one kilogramme a distance of one metre with a force of one *newton*. (The newton is defined as the force required to move one kilogramme with an acceleration of one metre per second per second.)

Example 1 An ideal gas of volume $0.1 \, m^3$ at a temperature of $273.15 \, K$ and a pressure of $10^5 \, N \, m^{-2}$ expands reversibly to three times its volume (a) at constant pressure and (b) at constant temperature. Find the work done in each case. ($R = 8.314 \, J \, K^{-1} \, mol^{-1}$ and $\ln N = 2.303 \log_{10} N$)

(a) At constant P, work done, $w = P(V_2 - V_1)$
$$= 10^5 \, (0.3 - 0.1)$$
$$= 2 \times 10^4 \, J$$

(b) In general, the work done is given by $w = \int_1^2 P dV$. This equation cannot be integrated directly since it is not known how the pressure varied during the expansion. However, substituting for P using the equation of state for an ideal gas ($PV = nRT$), we have, at constant T,

$$w = nRT \int_1^2 \frac{dV}{V} = nRT \ln \frac{V_2}{V_1} = P_1 V_1 \ln \frac{V_2}{V_1}$$

$$= 10^5 \times 0.1 \ln \frac{0.3}{0.1} = 10^5 \times 0.1 \times 2.303 \log_{10} 3$$

$$= 10\,980 \, J$$

Example 2 Calculate the work done when one mole of ideal gas at $300 \, K$ expands isothermally from $0.002 \, m^3$ to $0.005 \, m^3$ in the following ways: (1) at a finite rate of expansion against a constant opposing pressure of $10^5 \, N \, m^{-2}$; (2) quasi-statically and isothermally and (3) by expanding freely. ($R = 8.314 \, J \, K^{-1} \, mol^{-1}$)

(1)
$$w = \int_1^2 P_{opp} \, dV$$

At constant P,
$$w = P_{opp} \, (V_2 - V_1)$$
$$= 10^5 \, (0.005 - 0.002)$$
$$= 300 \, J$$

(2) Employing the equation derived in Example 1 for an isothermal reversible expansion,

$$w = nRT \ln \frac{V_2}{V_1}$$

$$= 8.314 \times 300 \times 2.303 \log_{10} \frac{5}{2}$$

$$= 2284 \text{ J}$$

(3) $P_{opp} = 0$, therefore, from Eq. (1.4), $w = 0$.

1.13   The $P-V$ diagram

The path of a process as shown on a graph of pressure P against volume V of a fluid is known as a $P-V$ diagram, an example being Figure 1.5.

Since

$$w = \int_A^B P dV$$

the area under the curve in the $P-V$ diagram is equal to the work done in the expansion from state A to state B. Therefore the work

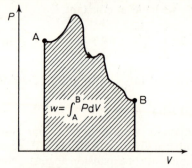

Figure 1.5. P–V diagram showing work done during a reversible change

done in passing from one state to another depends on the path chosen (see Figure 1.6). Different paths are possible because there is no restriction made on the temperature of the system.

Consider that the expansion from state A to B of a fluid takes place via path I. Now let us compress the fluid back to state A via path II (Figure 1.7). The process of going from A to B via one path and returning via another is known as a *cyclic process*. In general, a cyclic process is any succession of changes the result of which is to return the system to its original state.

According to our sign convention, the expansion work represented by the area under the curve *I* is positive and the compression work represented by the area under the curve *II* is negative. The net work

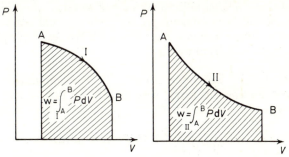

Figure 1.6. Showing the dependence of the work done on the path

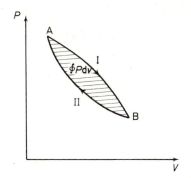

Figure 1.7. The work done during a cycle

done by the system in going once round the cycle is therefore represented by the shaded area enclosed by the curves *I* and *II*.

Net work done,

$$w = \int_{\substack{A \\ I}}^{B} PdV + \int_{\substack{B \\ II}}^{A} PdV$$

$$w = \oint PdV \qquad (1.9)$$

(*Note*: $\oint$ means integrated over the cycle.)

That is, in going once round the cycle, the system has done an amount of work given by Eq. (1.9) and represented by the shaded area. The energy to do this work has come from the heat absorbed by the fluid from its surroundings.

1.14   Work done by systems in reversible processes

Simple systems   The equilibrium states of systems which may be given by the functions of state pressure, volume and temperature are known as simple systems. The work done in a reversible process for such a system is given by Eq. (1.6).

*Wire   Consider a wire under a tension T. The infinitesimal work done, đw, to increase the length reversibly by dL is

$$đw = -T\,dL \tag{1.10}$$

The minus sign indicates that for an increase in length, i.e. positive dL, work is done *on* the system.

If a process is performed quasistatically, then at all times the tension in the wire is equal to the external forces, and, hence,

$$w = -\int_1^2 T\,dL$$

If the process is performed irreversibly, the tension will vary throughout the wire and will also depend on time. Hence, Eq. (1.10) cannot be employed.

*Surface film   The work done to increase reversibly the surface area of a liquid by dA, the liquid having surface tension θ, is given by

$$đw = -\theta\,dA$$

*Magnetisation   The work done to increase reversibly the magnetisation of a material by an amount dM in a magnetic field of intensity H is given by

$$đw = -H\,dM$$

REVERSIBLE CELLS

A galvanic cell employs a chemical or physical change to obtain electrical energy. A typical cell is the Daniell cell (Fig. 1.8). A zinc rod is immersed in zinc sulphate solution of concentration C_1, and a copper rod is immersed in copper sulphate solution of concentration C_2. The two solutions (electrolytes) $ZnSO_4$ and $CuSO_4$ are in electrical contact by means of the porous partition shown.

The cell may be written as

$$Zn \mid ZnSO_4(C_1) \mid CuSO_4(C_2) \mid Cu$$

where the vertical lines represent phase boundaries.

On connecting the zinc and copper electrodes by means of a wire, an electric current flows spontaneously. The potential difference between the electrodes when no current flows is known as the electromotive force (e.m.f.). The reaction occurring in the cell is

$$Zn + CuSO_4 \rightarrow ZnSO_4 + Cu \qquad (1.11)$$

The energy from the reaction which would normally be released as

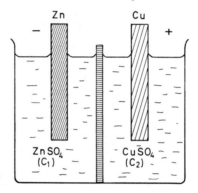

Figure 1.8. The Daniell cell

heat is converted into electricity. For a cell to be known as reversible the following three conditions must be fulfilled:

(1) if the e.m.f. of the cell is exactly balanced by an opposing potential difference, no reaction takes place;

(2) if an infinitesimal current is discharged by the cell, the reaction takes place – for the Daniell cell, for example, the reaction given by Eq. (1.11);

(3) if an infinitesimal current is passed through the cell in the opposite direction, the reverse reaction occurs – for the Daniell

$$Cu + ZnSO_4 \rightarrow CuSO_4 + Zn$$

The cell can only be employed reversibly when an infinitesimal current is involved, i.e. when the e.m.f. of the cell is only infinitesimally different from an opposing potential difference. If a finite current were taken from the cell, then concentration differences would be set up in the electrolytes around the electrodes. Also, if a finite current I passes through the cell, heat is generated equal to I^2R, where R is the total resistance of the circuit. This is known as Joulean heat and is analogous to energy being lost by friction.

Consider a reversible cell of e.m.f. E connected to a potentiometer (Figure 1.9). The balance point corresponds to no current flowing through the cell. When infinitesimally off-balance, a quantity of electricity dQ flows through the cell in time dt. The work done by the system is

$$\bar{d}w = EdQ \tag{1.12}$$

Since $dQ = Idt$, where I is the current flowing through the cell,

$$\bar{d}w = EIdt \tag{1.13}$$

During a reversible process, the work done is given by

$$w = \int_1^2 EdQ$$

so that
$$w = E(Q_2 - Q_1)$$

The charge of one mole of proton is known as *Faraday's constant*

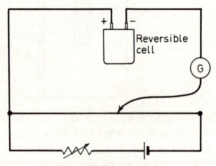

Figure 1.9. Potentiometer

$F = Le$, where L is Avogadro's constant and e is the charge of a proton. $F = 9.648\ 70 \times 10^4\ \text{C mol}^{-1}$ (coulombs per mole).

If an amount of electricity nF is passed through the cell (where n denotes the number of faradays passed through the solution) — that is, if

$$Q_2 - Q_1 = nF$$

then the work done is
$$w = nFE \tag{1.14}$$

WORK DONE BY VARIOUS SYSTEMS

All expressions for work are given by the product of an extensive property and an intensive property.

System	Extensive property	Intensive property	Reversible work done
mechanical	displacement (m)	force (newtons, N)	$F\mathrm{d}x$
simple	volume (m³)	pressure (N m⁻²)	$P\mathrm{d}V$
electrical cell	charge (coulombs, C)	potential difference (volts, V)	$E\mathrm{d}Q$
wire	distance (m)	tension (N)	$-T\mathrm{d}L$
surface changes	area (m²)	surface tension (N m⁻¹)	$-\theta\mathrm{d}A$
magnetic	magnetisation	intensity	$-H\mathrm{d}M$

1.15 Exact and inexact differentials

If a fluid undergoes a reversible change from state 1 to state 2, the work done is given by

$$w = \int_1^2 P\mathrm{d}V \qquad (1.15)$$

Even though the initial and final states are known, Eq. (1.15) cannot be integrated because it is not known how the pressure varied during the change. In other words, we do not know the path taken by the system in passing from state 1 to state 2. We have already seen that the work done is dependent on the particular path chosen. The differentials of parameters which depend on the path between states are known as *inexact differentials*, e.g. đ*w*, the line through the differential indicating an *inexact* differential. Hence, ∫đ*w* cannot be evaluated without information regarding the path taken.

The differentials of *functions of state* are known as *exact differentials*, e.g. d*V*, d*P*, d*T*. Exact differentials may be integrated since we are only concerned with the initial and final states, not with the path taken. The integral over a complete cycle of a function of state is zero, e.g.

$$\oint \mathrm{d}V = 0 \qquad (1.16)$$

whereas the integral of an inexact differential over a cycle is not zero (see Figure 1.7).

Equation (1.16) may be used to test whether a parameter is a function of state.

1.16 Partial differentials

In the derivation of thermodynamic relationships, it is often useful to employ some of the general rules of partial differential calculus.

Consider a variable z which may be determined from a knowledge of two other independent variables x and y ('independent' means that x cannot be found from a knowledge of y alone). Then z is a function of x and y, i.e.

$$z = f(x, y) \tag{1.17}$$

e.g. pressure is a function of V and T, $P = f(V, T)$ and, in fact, for an ideal gas, $P = (nRT)/V$.

We may define the partial derivative of z with respect to x as

$$\left(\frac{\partial z}{\partial x}\right)_y = \lim_{(\text{change in } x) \to 0} \frac{\text{small change in } z, y \text{ being kept constant}}{\text{the corresponding change in } x}$$

where the subscript y denotes that y is kept constant.

$$\left(\frac{\partial z}{\partial x}\right)_y = \lim_{\delta x \to 0} \frac{f(x + \delta x, y) - f(x, y)}{\delta x} \tag{1.18}$$

Partial differentiation involves no new differentiation principles. Functions are differentiated considering the subscript variable to be constant.

If z is plotted against x and y, a surface is obtained (see Figure 1.10).

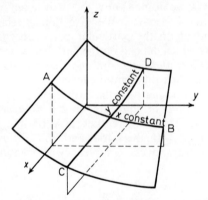

Figure 1.10.

The gradient at any point of the line AB is given by the partial differential $(\partial z/\partial y)_x$ and the gradient of the line CD by $(\partial z/\partial x)_y$.

The gradient of an isothermal P-V curve at any point is given by $(\partial P/\partial V)_T$ (see Figure 1.11).

It can be seen from the definition of the partial derivative (Eq. 1.18) that

$$\left(\frac{\partial z}{\partial x}\right)_y \cdot \left(\frac{\partial x}{\partial z}\right)_y = 1 \tag{1.19}$$

The differential of z is given by

$$dz = \left(\frac{\partial z}{\partial x}\right)_y \cdot dx + \left(\frac{\partial z}{\partial y}\right)_x \cdot dy$$ (1.20)

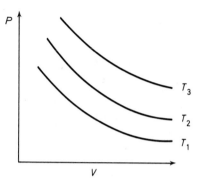

Figure 1.11. Isothermals on a P–V diagram

*PROOF OF EQUATION (1.20)

Consider that $z = f(x, y)$. The surface in Fig. 1.12 exhibits the dependence of z on x and y. The increase in z from the point A(x, y)

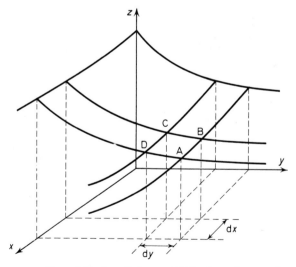

Figure 1.12. A small element ABCD of a surface

to $C(x+dx, y+dy)$ is equal to the increase in z from A to B plus the increase from B to C. On the line AB y is constant, and on the line CB x is constant.

The increase in z from A to B is

$$\left(\frac{\partial z}{\partial x}\right)_y . dx$$

The increase in z from B to C is

$$\left(\frac{\partial z}{\partial y}\right)_x . dy$$

Hence, the increase from A to C, dz, is given by

$$dz = \left(\frac{\partial z}{\partial x}\right)_y . dx + \left(\frac{\partial z}{\partial y}\right)_x . dy$$

THE CYCLIC RULE

Another important relation may be obtained from Eq. (1.20). Dividing by dy while keeping z constant,

$$0 = \left(\frac{\partial z}{\partial x}\right)_y . \left(\frac{\partial x}{\partial y}\right)_z + \left(\frac{\partial z}{\partial y}\right)_x$$

Rearranging,

$$\left(\frac{\partial z}{\partial x}\right)_y . \left(\frac{\partial x}{\partial y}\right)_z . \left(\frac{\partial y}{\partial z}\right)_x = -1 \qquad (1.21)$$

*1.17   Exact and inexact differentials

If, for a closed, one-component system, z is a function of state, then there must be an equation relating it and two other functions of state x and y:

$$z = f(x, y)$$

Therefore,

$$dz = \left(\frac{\partial z}{\partial x}\right)_y . dx + \left(\frac{\partial z}{\partial y}\right)_x . dy$$

or,

$$dz = M\,dx + N\,dy \qquad (1.22)$$

where $$M = \left(\frac{\partial z}{\partial x}\right)_y \qquad (1.23)$$

and $$N = \left(\frac{\partial z}{\partial y}\right)_x \qquad (1.24)$$

From Eq. (1.23), $$\left(\frac{\partial M}{\partial y}\right)_x = \frac{\partial^2 z}{\partial y \partial x} \qquad (1.25)$$

and from Eq. (1.24),

$$\left(\frac{\partial N}{\partial x}\right)_y = \frac{\partial^2 z}{\partial x \partial y} \qquad (1.26)$$

Since the order of partial differentiation is not important,

$$\boxed{\left(\frac{\partial M}{\partial y}\right)_x = \left(\frac{\partial N}{\partial x}\right)_y} \qquad (1.27)$$

This is known as *Euler's reciprocity relation* and is used as a test for an exact differential.

Example 1 Considering an ideal gas, show that volume is a function of state.
If V is a function of state, $\qquad V = f(T, P)$
For 1 mole of an ideal gas,

$$v = \frac{RT}{P} \qquad (1.28)$$

According to Eq. (1.20),

$$dv = \left(\frac{\partial V}{\partial T}\right)_P dT + \left(\frac{\partial V}{\partial P}\right)_T dP \qquad (1.29)$$

Using Eq. (1.28) in Eq. (1.29), we obtain

$$dv = \frac{R}{P} \cdot dT + \left(\frac{-RT}{P^2}\right) dP \qquad (1.30)$$

Therefore $M = R/P$ and $N = -RT/P^2$. We see that

$$\left(\frac{\partial M}{\partial P}\right)_T = -\frac{R}{P^2} = \left(\frac{\partial N}{\partial T}\right)_P$$

so that Euler's reciprocity relation is satisfied. Therefore v is a function of state.

Example 2   Considering an ideal gas, is *w* a function of state?

From Eq. (1.3),

$$đw = P\,dV \tag{1.31}$$

Substituting Eq. (1.30) into Eq. (1.31),

$$đw = P\left(\frac{R\,dT}{P} - \frac{RT\,dP}{P^2}\right)$$

$$= R\cdot dT - \frac{RT}{P}\cdot dP$$

In this case, $M = R$ and $N = -RT/P$

Hence, $$\left(\frac{\partial M}{\partial P}\right)_T = 0 \tag{1.32}$$

and $$\left(\frac{\partial N}{\partial T}\right)_P = -\frac{R}{P} \tag{1.33}$$

and Euler's reciprocity relation is not satisfied, meaning *w* is not a function of state. Therefore *w* cannot be written as a function of two functions of state, e.g. $w \neq f(P, V)$.

It is almost always possible to find an *integrating factor*, *I*, such that $$I\,đz = d\phi \tag{1.34}$$

where *z* is not a function of state but $d\phi$ is an exact differential. We must choose *I* so that Euler's relation is true,

i.e. $$\frac{\partial(MI)}{\partial y} = \frac{\partial(NI)}{\partial x}$$

*1.18   Simple systems

Simple systems are those for which the state may be given in terms of mass, pressure, temperature and volume alone.

THERMAL EXPANSIVITY

From elementary physics, the *coefficient of volume expansion* is defined by

$$\alpha = \frac{\text{change in volume per unit volume}}{\text{change in temperature}}$$

In differential terms, we have

$$\alpha = \frac{1}{V} \cdot \left(\frac{\partial V}{\partial T}\right)_P \tag{1.35}$$

α is also known as the *isobaric thermal expansivity*.

ISOTHERMAL COMPRESSIBILITY

Isothermal compressibility is defined by

$$\kappa = -\ \frac{\text{change in volume per unit volume}}{\text{change in pressure}}$$

For all known substances, an increase in pressure produces a a decrease in volume. Hence, the minus sign is introduced so that compressibility is a positive quantity.

For an infinitesimal change in pressure,

$$\kappa = -\ \frac{1}{V}\left(\frac{\partial V}{\partial P}\right)_T \tag{1.36}$$

The reciprocal of the isothermal compressibility is known as the *isothermal bulk modulus, β*

$$\beta = -V\left(\frac{\partial P}{\partial V}\right)_T \tag{1.37}$$

WORK DONE DURING A REVERSIBLE ISOTHERMAL CHANGE FOR A SIMPLE SYSTEM

The work done is given by

$$w = \int P\,\mathrm{d}V \tag{1.38}$$

Now V may be expressed in terms of T and P, i.e.

$$V = f(T, P)$$

Hence, $$\mathrm{d}V = \left(\frac{\partial V}{\partial T}\right)_P \mathrm{d}T + \left(\frac{\partial V}{\partial P}\right)_T \mathrm{d}P$$

For an isothermal process, $\mathrm{d}T = 0$; therefore we have

$$\mathrm{d}V = \left(\frac{\partial V}{\partial P}\right)_T \mathrm{d}P = -V\kappa\,\mathrm{d}P$$

Substituting into Eq. (1.38), we have that the reversible isothermal work done is

$$w = -\int PV\kappa\, dP \qquad (1.39)$$

For an *ideal gas*, $PV = nRT$ or $V = nRT/P$.

Hence,
$$\left(\frac{\partial V}{\partial P}\right)_T = -\frac{nRT}{P^2}$$

From Eq. (1.39), the reversible isothermal work done by an ideal gas between the states 1 and 2 is therefore

$$w = -nRT\int_1^2 \frac{dP}{P}$$

$$w = nRT\ln\frac{P_1}{P_2} \qquad (1.40)$$

Equation (1.39) applies to *solids* and *liquids*. However, to a good approximation, for solids and liquids V and κ remain constant except at very high pressures. Hence,

$$w = -V\kappa\int P\, dP$$

For a change from state 1 to state 2,

$$w = -\frac{V\kappa}{2}(P_2^2 - P_1^2) \qquad (1.41)$$

Problems

1.   Calculate the work done when 1 kg of water is converted into steam at atmospheric pressure ($101\,325\ \text{Nm}^{-2}$), the volume occupied by the steam being $1.669\ \text{m}^3$ and the molar volume of liquid water, $18 \times 10^{-6}\ \text{m}^3\,\text{mol}^{-1}$.
Answer $1.69 \times 10^5\ \text{J}$

2.   Calculate the work done when 2 moles of an ideal gas are iso-thermally and quasi-statically compressed from a volume of $2 \times 10^{-3}\ \text{m}^3$ to $10^{-3}\ \text{m}^3$ at 273 K, $R = 8.314\ \text{JK}^{-1}\,\text{mol}^{-1}$.
Answer $-3.147\ \text{kJ}$

3.   Derive expressions for the isobaric thermal expansivity (coefficient of thermal expansion) α and the isothermal compressibility κ

for a gas obeying (a) van der Waals' equation of state and (b) $P(v-b) = RT.$

Answer   (a)

$$\alpha = T^{-1}\left(1-\frac{2a}{RTv}+\frac{Pb}{RT}+\frac{3ab}{RTv^2}\right)^{-1}$$

$$\kappa = P^{-1}\left(1-\frac{b}{v}\right)\left(1-\frac{2a}{Pv^2}+\frac{2ab}{Pv^3}\right)^{-1}$$

(b)

$$\alpha = \frac{R}{Pv}$$

$$\kappa = P^{-1}\left(1-\frac{b}{v}\right)$$

4.   A fluid is contained in a cylinder of volume 1 m^3 at 101 325 Nm^{-2} and 300 K. If the pressure is increased to 10·132 MN m^{-2} and the compressibility of the fluid is 5×10^{-10} m^2 N^{-1}, calculate (a) the work done and (b) the change in volume.

Answer   (a) -250 J, (b) 5×10^{-5} m^3

2
The First Law of Thermodynamics

2.1 Brief historical introduction

From observations on the heat evolved during cannon boring, Count Rumford in 1798 suggested that heat was produced by mechanical work. The Caloric Theory, which was in favour at the time, postulated that heat was a substance — caloric — and that the heat flow from a hot body to a cold body was explained by the flow of caloric fluid. To account for Count Rumford's observations, the supporters of the Caloric Theory suggested that the metal turnings which were produced during the boring had a lower specific heat capacity than the metal. Hence, heat was evolved during the boring. Count Rumford then used a blunt borer which did not produce turnings. He found that the same amount of heat was evolved for an equivalent amount of work to that done previously with the sharp borer. Supporters of the Caloric Theory then said that heat was produced by the action of air on metal surfaces. In 1799, Sir Humphrey Davy rubbed two pieces of ice together in a vacuum and showed that they melted. However, this experiment was considered inconclusive.

Around 1842, James Joule was conducting experiments on the measurement of the mechanical equivalent of heat. At about this time, the law of conservation of energy was being generally accepted: *energy can neither be created nor destroyed, although it may be converted from one form to another.* Or, in other words, the energy of an isolated system remains constant. Energy is defined as the capacity to do work. Heat is merely another form of energy and must therefore obey the laws of conservation of energy.

2.2 Heat

We all know that heat flows from a hot body to a cold body when the two are in thermal contact. No further flow of heat will occur once the two bodies reach thermal equilibrium. Heat is energy that is in transit, i.e. in the process of being conveyed from one point to another. It is wrong to use such phrases as 'the heat *in* a body' or 'the heat possessed by a body'. Such phrases imply that heat is a static form of energy and hence that heat is a property of the system. Heat and work are merely ways in which we can alter the energy of the system. We can prove quite simply that heat and work are not functions of state by an experiment showing that heat and work depend on the path between the initial and final states of the system.

Figure 2.1. A system on which work may be done by a falling weight turning a paddle wheel

Consider an experiment in which water surrounded by an adiabatic wall may be stirred by a paddle wheel. The work done on the system results in a temperature rise, and this amount of work may be found from the distance the weight falls. The adiabatic walls are then removed and the system is placed in thermal contact with a water bath. In this way the system is brought back to its initial state, thus completing a cycle.

The heat transferred may be found by measuring the temperature change of the system. During the cycle, the only work done is that due to the falling weight, and the only heat transfer is that involving the water bath. It is therefore seen from this experiment that during this cyclic process the total work or heat involved is not zero,

i.e. $$\oint \dd w \neq 0 \text{ and } \oint \dd q \neq 0$$

Therefore, heat and work are not functions of state (see Section 1.15).

2.3 Formulation of the first law

Joule's experiments showed that, *for a system during a cyclic process, the sum of the work transferred is equal to the sum of the heat transferred* (heat and work being measured in the same units, e.g. joules),

$$\oint \dd w = \oint \dd q \qquad (2.1)$$

Rearranging, $$\oint (\dd q - \dd w) = 0 \qquad (2.2)$$

This means that, although heat and work are not functions of state, their difference is a function of state since the integral over the cycle is always zero (see Section 1.15).

Let $$dU = đq - đw \tag{2.3}$$

that is, $$\oint dU = 0 \tag{2.4}$$

The *function of state U* is known as *the internal energy* of the system.

The *first law of thermodynamics* states that *if heat đq is added to a system so as to increase the internal energy U by dU and also give rise to an amount of work đw, then*

$$đq = dU + đw \tag{2.5}$$

This differential form of the first law is a result of Eq. (2.2). In Eq. (2.5) we must use the same units for heat, energy and work.

The first law is merely a statement of the law of conservation of energy. Some of the heat given to the system increases the energy of the system and the remainder enables work to be done on the surroundings (Figure 2.2).

The first law applies to any process whether reversible or irreversible. Consider a finite process from state 1 to state 2 in which a

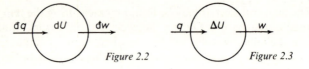

Figure 2.2   Figure 2.3

system absorbs heat q, does work w and its internal energy changes by ΔU from U_1 to U_2. From the law of conservation of energy,

$$q = \Delta U + w \tag{2.6}$$

where $$\Delta U = U_2 - U_1$$

Equation (2.6) is a statement of the first law as applied to a finite process.

SIGN CONVENTION

Note that in Eqs (2.5) and (2.6)

> heat *absorbed by* a system is considered positive,
> heat *rejected by* a system is considered negative,
> work done *by* a system is considered positive,
> work done *on* a system is considered negative.

There is no real need to remember the sign convention if the diagram of Figure 2.2 is kept in mind, always making certain that the law of conservation of energy holds.

Rearranging Eq. (2.6), we have:

$$\Delta U = q - w \tag{2.7}$$

Equation (2.7) defines the function of state U except for an arbitrary constant, since only the difference in U is involved.

2.4   Internal energy

The common definition of energy is the capacity to do work. *Internal energy* is merely the *energy content of the system*. Internal energy is easier to understand on the microscopic scale, where a system consists of particles (molecules or atoms) which have various forms of energy associated with them. The particles have velocities and therefore translational energy. Other forms of energy such as those due to rotation, vibration and interaction may be present. Electronic and nuclear energies also exist. Exactly which energies we consider the internal energy of our system to be comprised of will depend on our problem. This is permissible because we are only ever interested in changes in internal energy.

Internal energy must be an extensive property since, if we halve a system, we halve the energy.

Internal energy U is a function of state; hence, U is a function of any two other independent functions of state (say P, V) for a closed, one-component system:

$$U = f(P, V)$$

and therefore, $dU = \left(\dfrac{\partial U}{\partial P}\right)_V dP + \left(\dfrac{\partial U}{\partial V}\right)_P dV$

CONSERVATION OF ENERGY: PERPETUAL MOTION

Consider a cycle in which a system originally in state 1 is allowed to change to state 2 via path I and return to its original state via path II.

Then, from Eq. (2.4),

$$\int_{I_1}^{2} dU + \int_{II_2}^{1} dU = \oint_{I-II} dU = 0 \tag{2.8}$$

where $\displaystyle\int_{I_1}^{2}$ denotes integration from state 1 to state 2 via path I.

Now consider the cycle in which the system changes from state 1 to state 2 via path III and is then returned to state 1 via path II. Then,

$$\int_{III}\!\!{}_1^2 dU + \int_{II}\!\!{}_2^1 dU = \oint_{III-II} dU = 0 \qquad (2.9)$$

Hence, from Eqs (2.8) and (2.9),

$$\int_{I}\!\!{}_1^2 dU = \int_{III}\!\!{}_1^2 dU$$

That is, the change in energy from state 1 to state 2 via path I is equal to that via path III. If the energy change via path I had been different from that via path III, it would be possible to devise a cycle, I–III or III–I, resulting in a net gain in energy at its completion. It would therefore be possible to construct a perpetual-motion

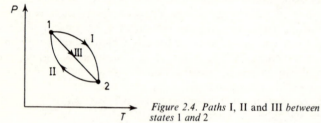

Figure 2.4. *Paths* I, II *and* III *between states* 1 *and* 2

machine of the first kind, this being a device which, without external interaction, does work continuously. This is, of course, contrary to the law of conservation of energy.

2.5   Reversible processes

Considering a simple system and the work done during volume changes, (Eq. (1.6)) the first law as applied to a reversible process is

$$\text{đ}q = dU + PdV \qquad (2.10)$$

This equation may be extended to include other kinds of reversible work which may be performed by a more complex system;

$$\text{đ}q = dU + PdV - TdL - \theta dA - HdM$$

2.6   Adiabatic processes

In adiabatic processes, no heat is allowed to enter or leave the system, $đq = 0$. Hence, from the first law,

$$\boxed{đw = -dU}$$
(2.11)

That is, the work done by the system is equal to the decrease in its internal energy.

Example   Calculate the increase in the internal energy when 1 mole of water at a pressure of 101 325 N m^{-2} and temperature 273·15 K is converted to steam at the same temperature and pressure. The latent heat of vaporisation of water is 2260 kJ kg^{-1} and the molar volume of an ideal gas is $2·24 \times 10^{-2}$ m^3 mol^{-1} and of liquid water is 18×10^{-6} m^3 mol^{-1} at 273·15 K and 101 325 N m^{-2}.

   Since the pressure remains constant, the work done by the steam in expanding is given by

$$w = P(V_2 - V_1)$$
$$= 101\ 325\ (2·24 \times 10^{-2} - 18 \times 10^{-6})$$
$$= 2267\ \text{J}$$

The heat absorbed by the water is (molecular weight of water being 18)

$$q = 18 \times 2260 = 40\ 680\ \text{J}$$

From the first law, $q = \Delta U + w$

$$\Delta U = q - w$$
$$= 38\ 413\ \text{J}$$

2.7   Heat changes at constant pressure: enthalpy

The differential form of the first law for reversible processes (Eq. 2.10) states

$$đq = dU + PdV$$

Considering a system that passes from state 1 to state 2, we have

$$\int_1^2 đq = \int_1^2 dU + \int_1^2 PdV$$
(2.12)

The left-hand side of Eq. (2.12) cannot be integrated unless the path between the states is known. The same applies to the PV term. If we consider that a system absorbs heat isobarically, i.e. at constant pressure, then the path is known. At constant P,

$$\int_1^2 đq = \int_1^2 dU + P \int_1^2 dV \qquad (2.13)$$

Integrating,

$$q = U_2 - U_1 + P(V_2 - V_1) \qquad (2.14)$$

where q is the heat absorbed or emitted at constant pressure. Rearranging Eq. (2.14),

$$q = U_2 + PV_2 - (U_1 + PV_1)$$

Defining a *new* function, *enthalpy*, H, by

$$\boxed{H = U + PV} \qquad (2.15)$$

we have

$$q = H_2 - H_1$$

or

$$\boxed{q = \Delta H} \qquad (2.16)$$

Note that this equation holds for any reversible isobaric process.

The definition (Eq. (2.15)) is not much help in enabling us to understand what is meant by enthalpy. This is hardly surprising since the definition has been made purely for mathematical convenience. In order to obtain a better understanding, we should examine Eq. (2.16), since we shall always be concerned with *enthalpy changes*. Equation (2.16) states that *the heat absorbed or emitted by a system during a reversible constant-pressure process is equal to the enthalpy change* from the initial to the final state.

Since the definition of enthalpy (Eq. (2.15)) involves the sum and product of functions of state (U, P and V), enthalpy must also be a function of state. We can see from Eq. (2.15) that enthalpy is an extensive property simply because U and V are extensive properties. The molar enthalpy, h, is given by

$$h = u + Pv$$

where u and v are the molar internal energy and molar volume, respectively. Molar enthalpy is an intensive property. Obviously, the enthalpy of a system containing n moles is nh.

A SIMPLER DERIVATION OF EQUATION (2.16)

From the first law, for a reversible process, we have

$$đq = dU + PdV$$

At constant P, $\quad\quad\quad\quad\quad \text{d}q = \text{d}U + \text{d}(PV)$

Hence, $\quad\quad\quad\quad\quad\quad \text{d}q = \text{d}(U + PV)$

Defining enthalpy as in Eq. (2.15), we have that, at constant P,

$$\boxed{\text{d}q = \text{d}H} \quad\quad\quad (2.17)$$

2.8 Heat changes at constant volume

If the volume of a simple system does not alter, then, from Eq. (1.3), no work can be done. That is, at constant volume $dV = 0$ and therefore $\text{d}w = 0$. Using this in the first law (Eq. (2.5)), we have that, at constant V,

$$\boxed{\text{d}q = \text{d}U} \quad\quad\quad (2.18)$$

Hence, the *heat absorbed or emitted at constant volume by a simple system is equal to the change in internal energy*.

ISOTHERMAL PROCESSES

In an isothermal change, the temperature of a system is constant. Therefore

$$\text{d}T = 0$$

2.9 Heat capacity

The heat capacity C of a system is the heat required to raise the temperature of the system by one degree.

$$C = \frac{\text{amount of heat absorbed by the system}}{\text{temperature rise}}$$

Since the heat capacity of a system may vary with temperature, it is better to define heat capacity in differential terms. If heat $\text{d}q$ is absorbed by a system and the resulting temperature rise is $\text{d}T$, then the heat capacity is given by

$$C = \frac{\text{d}q}{\text{d}T} \quad\quad\quad (2.19)$$

The heat capacity of a system containing 1 mole is known as the *molar heat capacity, c*. The heat capacity, C, of a system containing n moles is therefore given by

$$C = nc$$

Specific heat capacity is the heat capacity per unit mass.

Since heat is not a function of state, the amount of heat $đq$ absorbed by a system will depend on the path of the process. Therefore, the heat capacity of a system will also depend on the path, and there is an infinite number of possible paths. We must, therefore, put some restriction on the way heat is added to the system. The simplest and most common restrictions are for the system to be kept at constant volume or constant pressure. The heat capacity at constant volume, C_V, is given by

$$C_V = \left(\frac{\partial q}{\partial T}\right)_V \tag{2.20}$$

Using Eq. (2.18), this becomes

$$\boxed{C_V = \left(\frac{\partial U}{\partial T}\right)_V} \tag{2.21}$$

Hence, at constant V,

$$dU = C_V\, dT \tag{2.22}$$

The heat capacity at constant pressure, C_P, is given by

$$C_P = \left(\frac{\partial q}{\partial T}\right)_P \tag{2.23}$$

Using Eq. (2.16) or (2.17), this becomes

$$\boxed{C_P = \left(\frac{\partial H}{\partial T}\right)_P} \tag{2.24}$$

Hence, at constant P,

$$dH = C_P\, dT \tag{2.25}$$

The molar heat capacity at constant pressure, c_P, is given by

$$c_P = \left(\frac{\partial h}{\partial T}\right)_P$$

The molar heat capacity at constant volume, c_V, is similarly defined.
For a solid or liquid, C_P is very nearly equal to C_V. C_P and C_V differ significantly only for gases.

2.10   The difference between C_P and C_V

If an amount of heat q is added to a system at constant volume and this heat is enough to raise the temperature one degree, then q is equal to C_V. For C_P, in addition to the heat q, we must give the system energy to expand its boundaries in order that the pressure remain constant. Therefore, the heat capacity at constant pressure is greater than that at constant volume.

Let us consider a system at constant pressure. The first law (Eq. (2.10)) states that

$$\text{đ}q = \mathrm{d}U + P\,\mathrm{d}V \qquad (2.26)$$

Since internal energy U is a function of state, we may write

$$U = f(T, V)$$

and therefore

$$\mathrm{d}U = \left(\frac{\partial U}{\partial T}\right)_V \mathrm{d}T + \left(\frac{\partial U}{\partial V}\right)_T \mathrm{d}V \qquad (2.27)$$

Substituting for $\mathrm{d}U$ in Eq. (2.26),

$$\text{đ}q = \left(\frac{\partial U}{\partial T}\right)_V \mathrm{d}T + \left(\frac{\partial U}{\partial V}\right)_T \mathrm{d}V + P\,\mathrm{d}V \qquad (2.28)$$

Remembering that, at constant pressure, from Eqs. (2.16) or (2.17) and Eq. (2.25),

$$\text{đ}q = \mathrm{d}H = C_P\,\mathrm{d}T \qquad (2.29)$$

Equation (2.28) gives

$$C_P\,\mathrm{d}T = \left(\frac{\partial U}{\partial T}\right)_V \mathrm{d}T + \left[P + \left(\frac{\partial U}{\partial V}\right)_T\right]\mathrm{d}V$$

Dividing by $\mathrm{d}T$ throughout, at constant pressure, and using Eq. (2.21),

$$\boxed{C_P - C_V = \left[P + \left(\frac{\partial U}{\partial V}\right)_T\right]\left(\frac{\partial V}{\partial T}\right)_P} \qquad (2.30)$$

The $P\left(\dfrac{\partial V}{\partial T}\right)_P$ term is the work produced per degree rise in temperature at constant pressure. The $\left(\dfrac{\partial U}{\partial V}\right)_T\left(\dfrac{\partial V}{\partial T}\right)_P$ term is the work done against the internal cohesive or repulsive forces between the molecules. Comparing these two terms, we can see why $\left(\dfrac{\partial U}{\partial V}\right)_T$ is known as the internal pressure. It is the change in energy with volume at constant temperature. The internal pressure is large in solids and liquids but usually negligible in gases compared with the pressure P.

2.11   Joule's experiment

In an ideal gas, the internal pressure is zero since one of the conditions for ideality is that there are no cohesive forces between the

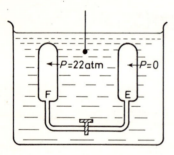

Figure 2.5. Joule's experiment

molecules. Joule attempted to measure the internal pressure of real gases. His apparatus consisted of two copper cylinders, F and E, immersed in a water-bath (Figure 2.5). The cylinder F was filled with dry air to a pressure of 22 atm (1 atm $= 101\,325$ Nm^{-2}) and E was evacuated. He measured the temperature of the bath with a thermometer accurate to 1/200th of a degree. He then opened a tap which allowed the air to flow into cylinder E. When the system had reached equilibrium, he measured the temperature again. He found *no* change in the initial and final temperatures of the water-bath.

Joule's experiment failed because his apparatus was not capable of detecting small enough temperature changes, the heat capacity of the water calorimeter being too large compared with the heat capacity of the air. Joule later continued his experiments in conjunction with Kelvin in an attempt to detect temperature changes as a result of expansion.

2.12 The Joule–Kelvin experiment

The experiment uses what is known as a *throttling process*. Gas in a container A is forced through a porous plug B to a container C (Fig. 2.6), the pressure in the container A being usually much higher

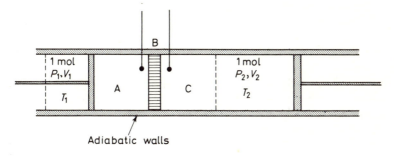

Figure 2.6. The Joule–Kelvin experiment

than that in C. The whole apparatus is thermally insulated. This experiment was successful in detecting a temperature difference either side of the plug B.

Let one mole of gas, pressure P_1 and volume V_1, in the container A be forced through the plug and let one mole of gas in the container C have a pressure P_2 and occupy a volume V_2.

The work done on the system to force one mole of gas (at constant pressure P_1) through the porous plug

= pressure × (final volume of gas in A − initial volume of gas in A)
= $P_1(0 - V_1) = -P_1 V_1$

The work done by the system due to the expansion (in container C)
= $P_2(V_2 - 0) = P_2 V_2$

The net work done is given by

$$w = P_2 V_2 - P_1 V_1 \qquad (2.31)$$

The first law states:

$$q = \Delta U + w \qquad (2.32)$$

Since the system is insulated, the process is adiabatic, i.e. no heat enters or leaves the system; $q = 0$. Let the internal energy of 1 mole of gas in A be U_1 and in C be U_2. Then, from Eqs (2.31) and (2.32), we have

$$0 = U_2 - U_1 + P_2 V_2 - P_1 V_1$$

$$U_2 + P_2 V_2 = U_1 + P_1 V_1$$

i.e. $$H_2 = H_1 \qquad (2.33)$$

This shows that the initial and final states of the throttling process have the same enthalpy.

The Joule–Kelvin coefficient is defined by

$$\mu_{\text{JK}} = \left(\frac{\partial T}{\partial P}\right)_H \qquad (2.34)$$

By measuring the temperature difference ΔT and the pressure difference ΔP between the gas in A and in C, the Joule–Kelvin coefficient is found. Since the pressure in A is always higher than that in C, it can be seen that if μ_{JK} is +ve, cooling occurs on expansion; if μ_{JK} is −ve, heating occurs on expansion.

For an ideal gas, $\mu_{\text{JK}} = 0$ because there are no internal forces between the molecules. The *inversion temperature* of a gas is the temperature at which neither cooling nor heating occurs on throttling. If a gas is above its inversion temperature, then a throttling process cannot be used to liquefy the gas. At room temperatures, all gases except hydrogen and helium show a cooling effect. For example, at room temperature and a pressure difference of 10^6 N m^{-2}, carbon dioxide cools by 10 K. The inversion point occurs where $\mu_{\text{JK}} = 0$. This is discussed in more detail in Chapter 9.

Example   Derive an expression for (a) the internal energy change in, and (b) the heat absorbed by, a system during an irreversible isobaric process from state 1 to state 2.

(a) The first law, which is true for any process, states

$$\text{d}q = \text{d}U + \text{d}w$$

Since internal energy is a function of state, it is independent of the path between two states and therefore independent of whether or not the process between the states is reversible. We may thus calculate the internal energy change considering the process to be reversible. Employing Eq. (2.10),

$$\text{d}q = \text{d}U + P\text{d}V$$

Employing Eq. (2.23), rearranging and integrating between states 1 and 2 at constant pressure,

$$\int_{U_1}^{U_2} \text{d}U = \int_{T_1}^{T_2} C_p\text{d}T - P\int_{V_1}^{V_2} \text{d}V$$

Therefore the internal energy change for an isobaric process is

$$\Delta U = \int_{T_1}^{T_2} C_p\text{d}T - P\Delta V \qquad (2.35)$$

(b) The heat absorbed for an irreversible isobaric process may be found by returning to the first law. Employing Eq. (1.3) and integrating,

$$q = \Delta U + \int_{V_1}^{V_2} P_{opp}\text{d}V$$

where P_{opp} is the pressure opposing volume change.
Using Eq. (2.35) to substitute for ΔU, we have that

$$q = \int_{T_1}^{T_2} C_p\text{d}T - P\Delta V + \int_{V_1}^{V_2} P_{opp}\text{d}V = \int_{T_1}^{T_2} C_p\,\text{d}T + (P_{opp} - P)\Delta V$$

2.13 Application of the first law to ideal gases

In thermodynamics, we shall say that a gas is perfect if the following statements are true:

(1) the equation of state for all values of P and T is

$$PV = nRT \qquad (2.36)$$

(n being the amount of gas, usually measured in moles);

(2) the internal energy is not a function of volume, i.e.

$$\left(\frac{\partial U}{\partial V}\right)_T = 0 \qquad (2.37)$$

Strictly speaking, to define an ideal gas all that we need is $PV = nRT$. The fact that there are no cohesive forces, and therefore that the internal energy does not alter with volume, is inherent in the equation of state. This will be shown in Section 9.2 (also see Section 5.15). We shall, for the present, assume that an ideal gas is defined by (1) *and* (2) although (1) is sufficient.

In general, $U = f(V, T)$ so that

$$dU = \left(\frac{\partial U}{\partial V}\right)_T dV + \left(\frac{\partial U}{\partial T}\right)_V dT \qquad (2.38)$$

Substituting from Eq. (2.21) and employing Eq. (2.37), we have

$$dU = C_V dT \qquad (2.39)$$

In fact, from (2) above we could have written

$$U = f(T). \qquad (2.40)$$

On integrating Eq. (2.39), we obtain

$$\Delta U = \int_{T_1}^{T_2} C_V dT \qquad (2.41)$$

Equation (2.39) is *only* true for processes involving an *ideal gas* and should not be confused with Eq. (2.22), which applies to *any* system but where the restriction of *constant volume* is essential.

Differentiating the equation of state of an ideal gas,

$$PdV + VdP = nRdT \qquad (2.42)$$

2.14   Difference of heat capacities (ideal gases)

From the first law we have, for a reversible process, using Eqs (2.10) and (2.39)

$$đq = C_V \, dT + P \, dV \qquad (2.43)$$

Hence, using Eq. (2.42),

$$đq = (C_V + nR)dT - VdP$$

We shall now consider that the system is kept at constant pressure, i.e. $dP = 0$. Then,

$$đq = (C_V + nR)dT$$

From Eq. (2.23),

$$C_P - C_V = nR \qquad (2.44)$$

This result could also have been obtained by employing the ideal-gas equation of state in Eq. (2.30) and remembering that

$$\left(\frac{\partial U}{\partial V}\right)_T = 0.$$

Differentiating Eq. (2.15),

$$dH = dU + PdV + VdP \qquad (2.45)$$

Employing Eqs (2.39) and (2.42), we obtain

$$dH = C_V dT + nRdT$$

Hence,

$$\boxed{dH = C_P dT} \qquad (2.46)$$

Note that for an ideal gas we have no restriction of constant P, unlike Eq. (2.25).

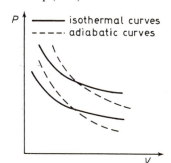

— isothermal curves

---- adiabatic curves

Figure 2.7. Isothermal and adiabatic curves

2.15   Reversible adiabatic processes (ideal gases)

For a reversible adiabatic process, we require that no heat enters or leaves the system during a reversible change. From the first law (Eq. 2.10), and since the process is adiabatic,

$$đq = dU + PdV = 0$$

so that

$$C_V dT + PdV = 0 \qquad (2.47)$$

Equation (2.42) states: $PdV + VdP = nRdT$

Using this to substitute for dT in Eq. (2.47),

$$C_V \left(\frac{PdV + VdP}{nR}\right) + PdV = 0$$

Since $nR = C_P - C_V$ we have

$$C_V PdV + C_V VdP + C_P PdV - C_V PdV = 0$$

Therefore
$$\frac{dP}{P} = -\frac{C_P}{C_V} \cdot \frac{dV}{V}$$

Let $C_P/C_V = \gamma$. Then   (2.48)

$$\int \frac{dP}{P} = -\int \frac{\gamma dV}{V}$$ (2.49)

In order to integrate the above equation, we must assume that γ is a constant. It has been found by experiment that, if the temperature change is not very large, this assumption is true for monatomic gases and very nearly so for polyatomic gases. For example, in the case of carbon monoxide, a temperature change of more than 2000 K is required to change the value of γ from 1.4 to 1.3. We may therefore write

$$\ln P + \gamma \ln V = \ln K$$

where $\ln K$ is the constant of integration. We hence obtain the familiar

$$\boxed{PV^{\gamma} = K}$$ (2.50)

This equation holds for *any* state of an ideal gas during a reversible adiabatic change.

Substituting for P with the aid of the equation of state of a perfect gas (Eq. 2.36), we have

$$\boxed{TV^{\gamma-1} = \text{constant}}$$ (2.51)

Similarly, we may obtain

$$\boxed{TP^{(1-\gamma)/\gamma} = \text{constant}}$$ (2.52)

Equations (2.50), (2.51) and (2.52) apply only to reversible changes. This may be seen by considering the free expansion of an ideal gas, as in Joule's experiment, where the temperature remains constant contrary to the prediction of Eqs (2.51) and (2.52).

2.16   Work done during a reversible adiabatic process (ideal gases)

We have already seen that the work done by a system during an adiabatic process is equal to the change in the internal energy (Eq. 2.11).

$$\text{d}w = -\text{d}U$$

From Eq. (2.39),
$$\boxed{\text{d}w = -C_V\text{d}T} \tag{2.53}$$

Assuming that C_V is temperature independent, we may integrate between T_1 and T_2,

$$w = -C_V\int_{T_1}^{T_2}\text{d}T$$

$$\boxed{w = -C_V(T_2 - T_1)} \tag{2.54}$$

*FURTHER CONSIDERATIONS

Consider a reversible adiabatic process such that a system passes from state 1 to state 2. We have shown (Eq. 1.6) that the work done by a system during an infinitesimal reversible change is given by

$$\text{d}w = P\text{d}V$$

Substituting for P from Eq. (2.50),

$$\text{d}w = KV^{-\gamma}\text{d}V$$

Integrating between state 1 and state 2,

$$w = K\int_{V_1}^{V_2}V^{-\gamma}\text{d}V$$

$$w = \frac{K}{1-\gamma}(V_2^{1-\gamma} - V_1^{1-\gamma})$$

But $P = KV^{-\gamma}$ and therefore $PV = KV^{1-\gamma}$. Hence,

$$\boxed{w = \frac{P_2V_2 - P_1V_1}{1-\gamma}} \tag{2.55}$$

Employing $PV = nRT$, we obtain

$$w = \frac{nR(T_2 - T_1)}{1 - \gamma} \tag{2.56}$$

2.17   **Work done during a reversible isothermal process (ideal gases)**

Consider a reversible isothermal process such that a system passes from state 1 to state 2. From the first law (Eq. 2.5),

$$đq = dU + đw$$

Since we are considering an isothermal process, $dT = 0$ and, hence, for a perfect gas we have, from Eq. (2.39), $dU = 0$. Therefore

$$đq = đw \tag{2.57}$$

That is, during an isothermal process, the heat absorbed by the system is equal to the work done by the system. Using Eq. (1.6),

$$đq = đw = PdV$$

Substituting for P, using $PV = nRT$ and then integrating between state 1 and state 2,

$$q = w = \int_1^2 \frac{nRTdV}{V}$$

Remembering that T is constant because the process is isothermal,

$$q = w = nRT \ln \frac{V_2}{V_1} \tag{2.58}$$

Using $PV = nRT$ to substitute for V in Eq. (2.58), we obtain another equation of the same form involving pressure;

$$q = w = nRT \ln \frac{P_1}{P_2} \tag{2.59}$$

Note that the ratio of the pressures is in the reverse order to that of the volumes in Eq. (2.58).

Example 1   Two moles of an ideal gas at $10^5 \, \text{Nm}^{-2}$ and 293 K are compressed to a pressure of $4 \times 10^5 \, \text{Nm}^{-2}$ (a) reversibly and isothermally, (b) reversibly and adiabatically. Find the work done in each case, given $R = 8.314 \, \text{JK}^{-1} \, \text{mol}^{-1}$, $\ln N = 2.303 \log_{10} N$ and the molar heat capacity $c_V = 3R/2$.

(a) For a reversible isothermal compression, we use Eq. (2.59):

$$q = w = nRT \ln \frac{P_1}{P_2}$$

$$q = w = 2 \times 8.314 \times 293 \ln \frac{10^5}{4 \times 10^5}$$

$$= -2 \times 8.314 \times 293 \times 2.303 \times 0.3979$$

$$= -4463 \, \text{J}$$

The minus sign indicates that the work is done *on* the system.

(b) Considering the reversible adiabatic compression, we have

$$\gamma = \frac{c_P}{c_V} = \frac{c_V + R}{c_V} = \frac{5}{3}$$

To find the final temperature, Eq. (2.52) is employed, i.e.

$$T_1 P_1^{(1-\gamma)/\gamma} = T_2 P_2^{(1-\gamma)/\gamma}$$

$$293 \times (10^5)^{-2/5} = T_2 (4 \times 10^5)^{-2/5}$$

Hence,
$$T_2 = 293 \times 4^{2/5}$$

$$T_2 = 405.2 \, \text{K}$$

The heat capacity at constant volume of the system is

$$C_V = nc_V = 2 \times 3R/2 = 3R.$$

The work done is given by

$$w = -C_V(T_2 - T_1)$$

$$= -3R(405.2 - 293)$$

$$= -2799 \, \text{J}$$

Example 2 One mole of an ideal gas at a temperature of 300 K expands adiabatically against a constant opposing pressure of $10^5 \, \text{N m}^{-2}$ from a volume of $10^{-4} \, \text{m}^3$ to $3 \times 10^{-4} \, \text{m}^3$. Calculate the final temperature, given $c_V = 16.0 \, \text{J K}^{-1} \, \text{mol}^{-1}$.

The irreversible work done is

$$w = \int_1^2 P_{\text{opp}} \, dV = P_{\text{opp}}(V_2 - V_1)$$

$$= 10^5(3 \times 10^{-4} - 10^{-4}) = 20 \, \text{J}$$

The work done during an infinitesimal adiabatic process for 1 mole of an ideal gas is

$$\mathrm{d}w = -\mathrm{d}u = -c_V \, \mathrm{d}T$$

$$w = -\int_1^2 c_V \, \mathrm{d}T$$

Since c_V is independent of T,

$$w = -c_V(T_2 - T_1)$$
$$= -16 \cdot 0(T_2 - 300) \, \mathrm{J}$$

Substituting the value for the work done as obtained above,

$$20 = -16 \cdot 0(T_2 - 300)$$

$$T_2 = 300 - \frac{20}{16 \cdot 0}$$

$$= 298 \cdot 7 \, \mathrm{K}$$

*2.18   Work done by a gas obeying van der Waals' equation

Van der Waals' equation of state of a gas does not reproduce the exact behaviour of a real gas, but it is a great improvement on the ideal-gas equation.

Van der Waals' equation for 1 mole of gas is

$$\left(P + \frac{a}{V^2}\right)(V - b) = RT \tag{2.60}$$

where a and b are constants, and, for n moles of gas,

$$\left(P + \frac{an^2}{V^2}\right)(V - nb) = nRT \tag{2.61}$$

This equation allows for the fact that (i) gases have a finite volume (b being related to this volume), (ii) intermolecular forces exist in the gas, decreasing the pressure expected if the gas were ideal (the term a/V^2 taking this into account).

Since intermolecular forces exist in a gas obeying van der Waals' equation,

$$\left(\frac{\partial U}{\partial V}\right)_T \neq 0$$

The work done in a reversible isothermal change from a state 1 to a state 2 is

$$w = \int_1^2 P \, dV$$

$$w = \int_1^2 \left(\frac{nRT}{V - nb} - \frac{an^2}{V^2} \right) dV$$

$$w = nRT \left[\ln (V - nb) \right]_{V_1}^{V_2} + \left[\frac{an^2}{V} \right]_{V_1}^{V_2}$$

$$w = nRT \ln \left(\frac{V_2 - nb}{V_1 - nb} \right) + \frac{an^2}{V_2} - \frac{an^2}{V_1} \qquad (2.62)$$

THE APPLICATION OF THE FIRST LAW TO THERMOCHEMISTRY

2.19 Heats of reaction at constant volume or constant pressure

The study of the heat changes accompanying chemical reactions, the formation of solutions and phase changes is known as thermochemistry. As we have seen, heat is not a function of state and therefore the heat changes accompanying reactions depend on the

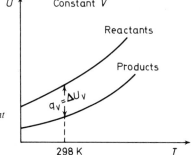

Figure 2.8. The heat of reaction at constant volume, q_V, is equal to the internal energy change, ΔU_V

external conditions of the system (system comprising reactants and products). The usual conditions are constant volume or constant pressure.

The reactants and products of a chemical reaction both have energies which may be associated with the forces binding the atoms forming them. The energy associated with the reactants may be greater than or less than that associated with the products, depending on the state and chemical nature of the reactants and products. The energy difference between the products and reactants at the initial temperature of a reaction is equal to the heat of reaction, this being denoted by q_V for a reaction carried out at constant volume and q_P for a reaction carried out at constant pressure. For example, the total internal energies of reactants and products at constant volume are plotted against temperature in Figure 2.8. q_V is the heat of reaction at constant volume at 298 K for an exothermic reaction (heat being liberated). From the first law, and using Eq. (2.18), we have, at constant volume,

$$q_V = \Delta U_V \qquad (2.63)$$

where subscript V indicates constant volume. That is, the heat of reaction at constant volume is equal to the change in the internal energy of a system.

Applying the first law, and using Eq. (2.16), we have, for constant pressure,

$$q_P = \Delta H_P \qquad (2.64)$$

where subscript P indicates constant pressure. That is, the heat of reaction at constant pressure is equal to the change in enthalpy of a system.

From Eq. (2.15), the heat change at constant pressure is given by

$$\Delta H_P = \Delta U_P + P\Delta V$$

Although the internal energy of a system may be considered as a function of T and P or T and V, the effect of isothermal pressure or volume changes in most practical cases may be neglected. In other words, from experiment it is found that the internal energy of a substance is independent of volume or pressure. This is, of course, exactly true for an ideal gas, but *not* for *real* gases because there exist intermolecular forces. At constant temperature the volumes and internal energies of *solids and liquids* are not altered significantly except by very large pressures. For a reaction carried out at a given temperature, we may say, as a good *approximation*, that

$$\Delta U_V = \Delta U_P$$

The difference between ΔU_V and ΔU_P can be shown by experiment to be negligible except when high pressures are involved. Therefore,

the heat of reaction at constant pressure is equal to the heat of reaction at constant volume plus the work done by the system when
the reaction proceeds at constant pressure to expand its boundaries:

$$q_P = q_V + P\Delta V \qquad (2.65)$$

Any further reference to enthalpy change or internal energy
change in this section will refer to the heats of reaction at constant
pressure and constant volume respectively. We shall therefore
omit the subscripts P and V and rewrite Eq. (2.65) as

$$\boxed{\Delta H = \Delta U + P\Delta V} \qquad (2.66)$$

If for a reaction there is no volume change, then, from Eq. (2.66),

$$\boxed{\Delta H = \Delta U} \qquad (2.67)$$

Consider

$$Cl_2(g) + H_2(g) \rightarrow 2HCl(g)$$

1 mole of chlorine reacts with 1 mole of hydrogen to give 2 moles
of hydrogen chloride.

1 volume of chlorine reacts with 1 volume of hydrogen to give 2
volumes of hydrogen chloride.

The symbols g, l and s are used to show the phase of a compound,
that is, gas, liquid or solid, respectively. If the reactants and products
are all in the solid or liquid phases, the change in volume involved
is usually negligible, and therefore Eq. (2.67) should be approximately
true. For a gaseous reaction, ΔV may be quite large.

The numbers preceding the chemical components in a balanced
chemical equation are known as *stoichiometric numbers* and are
proportional to the amounts of the chemical components that
change during a reaction — e.g., for the reaction given above, the
stoichiometric numbers for the reactants Cl_2 and H_2 are both 1,
and for the product HCl the number is 2.

Consider a reaction where the sum of the stoichiometric numbers
of the reactants is v_R and of the products is v_P. The difference is

$$v_P - v_R = \Delta v \qquad (2.68)$$

For the above reaction, $\Delta v = 2 - (1+1) = 0$. For

$$SO_2 + \tfrac{1}{2}O_2 \rightarrow SO_3$$

$$\Delta v = 1 - (1 + \tfrac{1}{2}) = -\tfrac{1}{2}$$

Let the volume occupied by 1 mole of gas, i.e. the molar volume, be v at the given temperature and pressure of the reaction. Then the change in volume when stoichiometric amounts react is

$$\Delta v = v\Delta v$$

Substituting this into Eq. (2.66),

$$\Delta H = \Delta U + Pv\Delta v \qquad (2.69)$$

Hence, assuming that the gases behave ideally,

$$\boxed{\Delta H = \Delta U + RT\Delta v} \qquad (2.70)$$

Chemical reactions which evolve heat are said to be *exothermic* and those which absorb heat are said to be *endothermic*.

An exothermic reaction is

$$H_2(g) + \tfrac{1}{2}O_2(g) \rightarrow H_2O(g)$$

$\Delta H = -241 \cdot 82$ kJ at 298 K and a pressure of 101 325 N m^{-2}.

In an *exothermic reaction*, heat is *emitted* from the system (reactants and products). Therefore, the heat evolved is considered to be *negative*, in order to be in agreement with the sign convention used with the first law.

Endothermic heats of reaction are considered to be *positive*, since the heat is *absorbed* by the system.

An endothermic reaction is

$$H_2O(g) \rightarrow H_2(g) + \tfrac{1}{2}O_2(g)$$

$\Delta H = 241 \cdot 82$ kJ at 298 K and a pressure of 101 325 N m^{-2}.

Example   Find the heat evolved at constant volume during the reaction forming 1 mole of water in the gaseous phase at 298 K from its elements. ($R = 8 \cdot 314$ J K^{-1} mol^{-1}.)

From above,

$$\Delta H = -241 \cdot 82 \text{ kJ at 298 K and 101 325 N m}^{-2}$$

$$\Delta v = 1 - (1 + \tfrac{1}{2}) = -\tfrac{1}{2}$$

Assuming that the reactants and products behave as ideal gases, from Eq. (2.70),

$$-241\,820 = \Delta U + 8\cdot314 \times 298 \times (-\tfrac{1}{2})$$

Hence, $\qquad \Delta U = -240\,581\text{ J} = -240\cdot581\text{ kJ}$

ENTHALPY OF FORMATION

The standard state of a substance is the state of its stable phase at 101 325 N m^{-2} (= 1 atm) *and the temperature considered,* e.g. at 101 325 N m^{-2} and 298 K, oxygen is gas, hydrogen is gas, mercury is liquid, sulphur is rhombic crystalline. It is usual to give heats of reaction as standard enthalpy changes $\Delta H^{\ominus}$ (298 K),

$$\tfrac{1}{2}H_2(g) + \tfrac{1}{2}Cl_2(g) = HCl(g); \Delta H^{\ominus}(298\text{ K}) = -92\cdot20\text{ kJ}$$

where the superscript $\ominus$ indicates that all substances are in the standard state and 298 K indicates the temperature considered.

By convention, *the enthalpies of elements in their standard state is zero.* The enthalpy change when 1 mole of compound is formed from its elements, the reactants and products being in their standard states, is known as the *standard enthalpy of formation,* $\Delta h^{\ominus}$ (298 K).

The reason for this convention is that in most everyday experiments there is conservation of atoms in a reaction. This is, of course, not true for nuclear reactions.

2.20 Hess' law

As a direct result of the first law, or the law of conservation of energy, enthalpy and internal energy are functions of state and therefore independent of the path of a process. This explains why Hess' law is true, this law stating: *the heat change accompanying a chemical reaction is the same whether the reaction takes place in one or more stages.*

Example $\quad C + \tfrac{1}{2}O_2 \rightarrow CO; \Delta H^{\ominus}$ (298 K) $= -111\cdot36\text{ kJ}$

$\qquad\qquad \underline{CO + \tfrac{1}{2}O_2 \rightarrow CO_2; \Delta H^{\ominus}\text{ (298 K)} = -283\cdot30\text{ kJ}}$

$\qquad\qquad C + O_2 \rightarrow CO_2; \Delta H^{\ominus}$ (298 K) $= -394\cdot66\text{ kJ}$

Hess' law may be employed to calculate heats of reactions which have not been determined experimentally.

The heats of reaction may also be found if the standard heats of formation of the reactants and products are known:

$$\Delta H^\ominus (298 \text{ K}) = \sum_{\text{products}} v\Delta h^\ominus (298 \text{ K}) - \sum_{\text{reactants}} v\Delta h^\ominus (298 \text{ K}) \quad (2.71)$$

where v is the corresponding stoichiometric number.

Example $CH_4(g) + 2O_2(g) \rightarrow CO_2(g) + 2H_2O(l)$

Given: substance $\Delta h^\ominus$ (298 K)
                     $CH_4(g)$ $-74.33 \text{ kJ mol}^{-1}$
                     $CO_2(g)$ $-393.20 \text{ kJ mol}^{-1}$
                     $H_2O(l)$ $-285.40 \text{ kJ mol}^{-1}$

$\Delta H^\ominus$ (298 K) $= [-393.20 + 2(-285.40)] - [-74.33 + 2(0)]$
                     $= -889.67 \text{ kJ}$

2.21   Measurement of heats of reaction (calorimetric methods)

Of all the possible chemical reactions that can occur, there are only a few for which the heat of reaction can be accurately determined. In order that accurate measurements can be made, we require that (1) the reaction is fast, or can be made fast by the addition of catalyst, so that the heat is absorbed or emitted over a short period of time; (2) the reaction goes to completion, so that we need not make corrections for unreacted substances; (3) there are no side-reactions, i.e. the reaction being studied is the only one that occurs.

Many methods have been used for the measurement of heats of reaction, some of which have been extremely ingenious. Calorimetry is a very large subject in its own right and we shall outline here two basic methods.

ISOTHERMAL CALORIMETER

The reaction process is carried out in the calorimeter B (Figure 2.9), which is surrounded by a water jacket A. This jacket is maintained at a given temperature by means of a thermostat. The temperature rise or fall in the calorimeter B due to the reaction is measured by a suitable thermometer. From the temperature change and the heat capacity of the calorimeter and contents, the heat of the reaction may be calculated.

The heat capacity of the calorimeter and its contents may be found by placing a heating coil in the calorimeter and finding the temperature rise for a known quantity of electricity. In this method it is usually advisable to make cooling corrections.

A slight modification of the procedure above is to find the exact quantity of electricity required to produce the same rise in tempera-

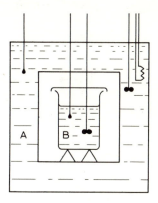

Figure 2.9. Calorimeter with surrounding jacket

ture as the reaction. Sometimes, in order to minimise heat losses, the calorimeter is placed in a Dewar flask.

The term isothermal implies that the environment jacket A of the calorimeter B is maintained at a constant temperature, usually to within 0·005 K. Since the pressure is constant during the reaction, the enthalpy change ΔH of the reaction is measured.

ADIABATIC CALORIMETER

This method aims at avoiding all heat losses. The temperature of the water jacket A is continuously adjusted in an attempt to make it at all times equal to that of the calorimeter. This can be done automatically by electronic devices to better than 0·1 K. Since the calorimeter and the water jacket are always at the same temperature, there is no heat exchanged between them, and therefore no cooling corrections need to be made. The enthalpy change is found by measuring the quantity of heat required to produce the same temperature rise as the reaction.

THE BOMB

The bomb is normally used to measure heats of combustion, experiments being carried out, as before, isothermally or adiabatically. The bomb is a very strong cylindrical vessel (Figure 2.10) which is placed in the calorimeter B of Figure 2.9. It is filled with oxygen to a pressure of 2–3 MN m^{-2} (20–30 atm). The weighed material, which is placed on a small cup, is ignited by passing a current through the ignition wire. The temperature rise of the calorimeter B is

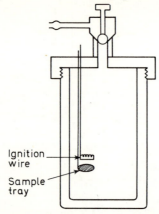

Ignition wire

Sample tray

Figure 2.10. Bomb calorimeter

measured, and allowance for the heat evolved by the ignition wire is made. Usually the bomb calorimeter is calibrated by using a substance whose heat of combustion is very accurately known, such as benzoic acid. Since the reaction occurs at constant volume, the bomb calorimeter measures the change in internal energy, ΔU, of a reaction.

2.22   Bond energies

The concept of bond energy enables estimations of heats of forma-tion or heats of reaction to be made. Bond energy is the average energy per mole required to break a given bond (see Table 2.1).

Table 2.1

Bond	Bond energy (kJ mol^{-1})(298 K)
H—H	430
Cl—Cl	238
C—C	335
H—Cl	425
O—H	464
C—H	410
C—Cl	326

The energy required to break a bond varies in different compounds. Bond energy is taken to be the average energy required to break a given bond, considering all possibilities.

Heats of reaction are obtained by considering the total energy required for the bonds broken in the reactants minus the total energy required for the bonds formed in the products.

Example 1 Find the heat of reaction for $H_2(g) + Cl_2(g) \rightarrow 2HCl(g)$
From Table 2.1,

energy of bonds broken: $\left.\begin{array}{l} H-H \ 430 \ kJ \ mol^{-1} \\ Cl-Cl \ 238 \ kJ \ mol^{-1} \end{array}\right\}$ total 668 kJ

energy of bonds formed: $2 \ H-Cl, \ 2(425) = 850 \ kJ$

Therefore heat of reaction, $\Delta H = 668 - 850 = -182 \ kJ$

Example 2 Find the heat of reaction for $CH_4 + Cl_2 \rightarrow CH_3Cl + HCl$
From Table 2.1,

energy of bonds broken: $\left.\begin{array}{l} C-H, \ 410 \ kJ \ mol^{-1} \\ Cl-Cl, \ 238 \ kJ \ mol^{-1} \end{array}\right\}$ total $= 648 \ kJ$

energy of bonds formed: $\left.\begin{array}{l} C-Cl, \ 326 \ kJ \ mol^{-1} \\ H-Cl, \ 425 \ kJ \ mol^{-1} \end{array}\right\}$ total $= 751 \ kJ^{-1}$

Therefore, $\Delta H = 648 - 751 = -103 \ kJ$

2.23   Temperature dependence of heats of reaction

We may know the heat of a reaction at one temperature and wish to find it at another. In Figure 2.11, the enthalpy of the reactants (upper curve) and of the products (lower curve) of an exothermic reaction are plotted against T. The standard enthalpy changes of the reaction, at temperatures 298 K and T_1, are shown.

We may note, in passing, that most published results are quoted at 298 K and $101 \ 325 \ N \ m^{-2}$ ($= 1 \ atm$), whereas older studies were referred to 291 K. Hence, caution must be exercised when comparing results.

The enthalpy change ΔH of a reaction may be written as

$$\Delta H = H_{products} - H_{reactants}$$

where $H_{products}$ is the sum of the enthalpies of the products, and similarly for $H_{reactants}$ (see Eq. 2.71).

Differentiating with respect to temperature at constant pressure,

$$\left(\frac{\partial \Delta H}{\partial T}\right)_P = \left(\frac{\partial H_{\text{products}}}{\partial T}\right)_P - \left(\frac{\partial H_{\text{reactants}}}{\partial T}\right)_P$$

From Eq. (2.24),

$$\left(\frac{\partial \Delta H}{\partial T}\right)_P = (C_P)_{\text{products}} - (C_P)_{\text{reactants}} = \Delta C_P$$

(To clarify the meaning of ΔC_P, see the examples below.)

$$\left(\frac{\partial \Delta H}{\partial T}\right)_P = \Delta C_P \tag{2.72}$$

Integrating (at constant pressure),

$$\Delta H_2 - \Delta H_1 = \int_{T_1}^{T_2} \Delta C_P \,. \, \mathrm{d}T \tag{2.73}$$

where ΔH_1 and ΔH_2 are the enthalpy changes of the reaction at temperatures T_1 and T_2.

That is, the difference in the heats of reaction at the temperatures T_1 and T_2, at constant pressure, is equal to the heat required to

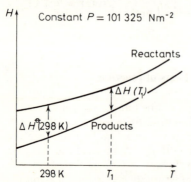

Figure 2.11. Enthalpy change of a reaction at different temperatures

raise the temperature of the reactants and products from T_1 to T_2, at constant pressure.

Equation (2.73) is known as *Kirchhoff's equation*.

In an exactly similar manner, it can be shown that

$$\Delta U_2 - \Delta U_1 = \int_{T_1}^{T_2} \Delta C_V \,. \, \mathrm{d}T \tag{2.74}$$

If we assume that ΔC_P is temperature independent, Eq. (2.73) may be integrated to give

$$\Delta H_2 - \Delta H_1 = \Delta C_P(T_2 - T_1) \qquad (2.75)$$

The temperature dependence of heat capacities (over a given temperature range) is sometimes expressed as a power series,

$$c_P = \alpha + \beta T + \gamma T^2 + \ldots \qquad (2.76)$$

Table 2.2 Molar heat capacities of gases at constant pressure,
$c_P = \alpha + \beta T + \gamma T^2$ J K^{-1} mol^{-1}

	α	$\beta \times 10^3$	$\gamma \times 10^7$
H_2	29·4	−0·836	20·6
O_2	25·7	12·9	−38·56
N_2	27·2	5·22	−0·042
Cl_2	31·6	10·1	−40·3
HCl	28·1	1·8	15·4
CO	26·8	6·93	−819
$H_2O(g)$	30·3	9·6	11·8
NH_3	33·6	2·93	213·0
CH_4	14·12	75·2	−179·8

Using this power series, Eq. (2.73) may be written as

$$\Delta H_2 - \Delta H_1 = \int_{T_1}^{T_2} (\Delta\alpha + \Delta\beta T + \Delta\gamma T^2 + \ldots) dT$$

so that we have

$$\Delta H_2 - \Delta H_1 = \Delta\alpha(T_2 - T_1) + \frac{\Delta\beta}{2}(T_2^2 - T_1^2) + \frac{\Delta\gamma}{3}(T_2^3 - T_1^3) \ldots$$

$$(2.77)$$

Example 1 The standard enthalpy of formation of liquid water at 298 K is $-285\cdot83$ kJ mol^{-1}. The molar heat capacities at a constant pressure of 101 325 N m^{-2} are:

$$c_P(H_2) = 29\cdot05 \text{ J K}^{-1} \text{ mol}^{-1}$$

$$c_P(O_2) = 25\cdot92 \text{ J K}^{-1} \text{ mol}^{-1}$$

$$c_P(H_2O(l)) = 75\cdot30 \text{ J K}^{-1} \text{ mol}^{-1}$$

Find the heat of reaction at 373 K.

The reaction is

$$H_2(g) + \tfrac{1}{2}O_2(g) \rightarrow H_2O(l); \quad \Delta H^\circ (298 \text{ K}) = -285{\cdot}83 \text{ kJ}$$

Since all the heat capacities are temperature independent, we can use Eq. (2.75):

$$\Delta c_P = (c_P)_{\text{products}} - (c_P)_{\text{reactants}}$$
$$= 75{\cdot}30 - (29{\cdot}05 + \tfrac{1}{2} \times 25{\cdot}92)$$
$$\Delta c_P = 33{\cdot}29 \text{ J K}^{-1} \text{ mol}^{-1}$$

Now

$$\Delta H_2 - \Delta H_1 = \Delta c_P(T_2 - T_1)$$

so that

$$\Delta H^\circ (373 \text{ K}) - (-285\ 830) = 33{\cdot}29\ (75)$$

Hence,

$$\Delta H^\circ (373 \text{ K}) = -285{\cdot}830 + 2{\cdot}496 = -283{\cdot}334 \text{ kJ}$$

Example 2   Calculate the enthalpy change at 398 K for the reaction

$$N_2 + 3H_2 \rightarrow 2NH_3; \quad \Delta H^\circ (298 \text{ K}) = -92{\cdot}41 \text{ kJ}$$

The molar heat capacities are (in $\text{J K}^{-1} \text{ mol}^{-1}$):

$$c_P(N_2) = 27{\cdot}26 + 5{\cdot}23 \times 10^{-3} T - 4{\cdot}18 \times 10^{-9} T^2$$
$$c_P(H_2) = 29{\cdot}02 - 8{\cdot}35 \times 10^{-4} T + 20{\cdot}80 \times 10^{-7} T^2$$
$$c_P(NH_3) = 25{\cdot}86 + 32{\cdot}94 \times 10^{-3} T - 30{\cdot}42 \times 10^{-7} T^2$$

We have that

$$\Delta c_P = 2c_P(NH_3) - [c_P(N_2) + 3c_P(H_2)]$$

Hence,

$$\Delta c_P = -62{\cdot}60 + 63{\cdot}06 \times 10^{-3} T - 123{\cdot}20 \times 10^{-7} T^2 \text{ J K}^{-1}$$

$$\Delta H_2 - \Delta H_1 = \int_{T_1}^{T_2} \Delta c_P \, dT$$

Therefore

$$\Delta H^\circ (398 \text{ K}) - (-92\ 410)$$

$$= \int_{298}^{398} (-62{\cdot}60 + 63{\cdot}06 \times 10^{-3} T - 123{\cdot}20 \times 10^{-7} T^2) \, dT$$

$$= \left[-62 \cdot 60T + 63 \cdot 06 \times 10^{-3} \frac{T^2}{2} - 123 \cdot 20 \times 10^{-7} \frac{T^3}{3} \right]_{298}^{398}$$

$$\Delta H^\ominus (398 \text{ K}) = -96 \cdot 50 \text{ kJ}$$

Problems

1. Calculate the maximum work obtainable from an isothermal expansion of one mole of an ideal gas from 10^{-3} m^3 to 10^{-2} m^3 at 298 K. $R = 8 \cdot 314 \text{ J K}^{-1} \text{ mol}^{-1}$.
Answer 5·7 kJ

2. Obtain an expression for $c_P - c_V$ for a van der Waals' gas.

Answer $c_P - c_V = R + R \left(\dfrac{2a}{Pv^2} - \dfrac{2ab}{Pv^3} \right) \left(1 - \dfrac{a}{Pv^2} + \dfrac{2ab}{Pv^3} \right)^{-1}$

3. Determine the work done during the isothermal reversible expansion from state 1 to state 2 of a gas obeying the equation of state $PV = A + BP + CP^2$.

Answer $$w = A \ln \frac{P_2}{P_1} + \frac{C}{2}(P_2^2 - P_1^2)$$

4. Calculate the enthalpy change of the reaction

$$C(s) + H_2O(g) \rightarrow CO(g) + H_2(g); \quad \Delta H^\ominus (298 \text{ K}) = 131 \cdot 2 \text{ kJ}$$

at 473 K, the molar heat capacities in $\text{J K}^{-1} \text{ mol}^{-1}$ being

$$c_P(C) = 11 \cdot 16 + 0 \cdot 010 \, 93 \, T$$
$$c_P(H_2O) = 29 \cdot 25 + 0 \cdot 011 \, 58 \, T + 11 \cdot 1 \times 10^{-7} \, T^2$$
$$c_P(CO) = 27 \cdot 6 + 0 \cdot 0510 \, T - 800 \times 10^{-7} \, T^2$$
$$c_P(H_2) = 27 \cdot 67 + 0 \cdot 003 \, 39 \, T - 21 \cdot 0 \times 10^{-7} \, T^2$$

Answer 133·2 kJ

5. Show that, at the intersection of an adiabatic and an isothermal curve on a P-V diagram for an ideal gas, the ratio of the slopes of the curves is C_P/C_V.

6. An ideal gas at 273 K, pressure $1\,013\,250 \text{ N m}^{-2}$ and volume 10^{-2} m^3 is allowed to expand reversibly to a pressure of 101 325 N m^{-2} (a) isothermally and (b) adiabatically. Calculate the work done in each case, given that one mole of an ideal gas at 101 325 N m^{-2} and 273 K occupies $2 \cdot 241 \times 10^{-2} \text{ m}^3 \text{ mol}^{-1}$, $c_V = 3R/2$ and $R = 8 \cdot 314 \text{ J K}^{-1} \text{ mol}^{-1}$.
Answer (a) 23·31 kJ, (b) 9·125 kJ

7.   One mole of an ideal gas at 300 K expands isothermally from a pressure of 1 000 000 N m^{-2} to 100 000 N m^{-2}. Calculate the work done if (a) the expansion is irreversible and frictionless, against a constant pressure of 10 000 N m^{-2}; (b) the expansion is reversible; (c) the expansion is against zero pressure.

Answer   (a) 225·3 J mol^{-1}, (b) 5762 J mol^{-1}, (c) 0

8.   Given that $C_6H_6(l) + 7\frac{1}{2}O_2(g) = 6CO_2 + 3H_2O(l)$; $\Delta H^\circ(298\ K) = -3265\ kJ$ calculate the heat of reaction at constant volume at 298 K, assuming the gaseous components to be ideal.

Answer -3260 kJ

9.   Show that for a van der Waals' gas undergoing an adiabatic reversible process,

$$T(v-b)^{R/c_V} = \text{constant}$$

10.   By graphical integration of the data given below, find the enthalpy change of silver when heated from 35 K to 100 K.

T/K	c_P (silver)/J mol^{-1}
35	6·65
43	9·33
45·5	10·15
51·5	11·80
77·0	17·00
100·0	19·73

Answer   955·9 J mol^{-1}

11.   Two moles of an ideal gas at 300 K and 10^6 N m^{-2} expand adiabatically to a pressure of 10^5 N m^{-2} (a) reversibly and (b) irreversibly against a constant pressure of 10^4 N m^{-2}. Given $c_V = 3R/2$, $R = 8·314$ J K^{-1} mol^{-1}, calculate the work done in each case.

Answer   (a) 4504 J, (b) 468·9 J

*12.   Find the work done when an ideal gas of volume 10^{-5} m^3 at 10^5 N m^{-2} expands reversibly and adiabatically to 2×10^{-5} m^3, given $\gamma = 5/3$.

Answer   0·561 J

3

The Second Law of Thermodynamics

3.1 Limitations of the first law

The first law *denies* the possibility of creating or destroying energy. This is equivalent to a denial of the idea that a machine may operate continuously, without interaction with its surroundings, to produce a continuous supply of energy (see Figure 3.1), i.e. this is a denial of what is known as perpetual motion of the first kind.

The first law only tells us that there must be a balance between the energy before and after a process. It is *not concerned with the direction* of the process. All naturally occurring processes are said to be *spontaneous processes* and take place without external influence. Not every process that is consistent with the law of conservation of energy is a spontaneous process.

Example 1 Consider a metal bar, hot at one end and cold at the other. Heat will flow *spontaneously from the hot end to the cold end* until the bar attains a uniform temperature. We never observe the reverse, i.e. a metal bar at a uniform temperature spontaneously becoming hotter at one end, even though this would not violate the first law since we still have an energy balance before and after the change.

Example 2 When a brake is applied to stop a wheel from rotating, the brake becomes hot owing to friction, and the internal energy of the brake increases. The reverse process is never observed – a hot brake losing its internal energy to a wheel which starts rotating – even though this process would not violate the first law.

Example 3 Consider a tank of water containing a paddle which will rotate when the weight W falls (Figure 3.2). The spontaneous fall of the weight decreases its potential energy and increases the internal energy of the water. The reverse process, i.e. the water cooling and the weight rising, is never observed, even though, once again, the first law would not be violated.

Figure 3.1. Perpetual motion of the first kind

Figure 3.2. Paddle wheel doing work on a system

Example 4 Zinc spontaneously dissolves in sulphuric acid, a definite amount of heat being liberated. The reverse process (electrolysis) is not spontaneous. Without external influence, the reaction proceeds in only one direction.

These examples indicate that *there exists a directional law which limits the way energy may be transferred or transformed.* This law is, in fact, the *second law of thermodynamics.*

The first law fails to explain why chemical reactions do not all proceed to completion. For example, consider the reaction $CO(g) + H_2O(g) \rightleftharpoons CO_2(g) + H_2(g)$. If originally 1 mole of CO and 1 mole of H_2O were in the reaction vessel, then, if the reaction went to completion, 1 mole of CO_2 and 1 mole of H_2 would be produced. If the reaction vessel is examined at equilibrium, the yield is less than 100%.

Before giving a statement of the second law, we shall consider some of our experiences with heat and work, since the laws of thermodynamics, like any other of the laws of science, are laws derived from experience.

3.2   Clarification of some terms

A machine that can convert heat into work is known as a *heat engine,* for example, the steam engine, the internal combustion engine.

The material comprising the system which does the work is known as the *working substance,* for example, steam, petrol–air mixture. Due to the nature of the steam and the internal combustion engines, the working substance is rejected after each cycle and a new quantity of working substance introduced. In an ideal engine, this is not necessary.

A heat engine is said to *operate in a cycle* if the working substance is returned to its original state at the end of a cycle.

In an *ideal* engine, friction and all similar means of energy dissipation are absent, e.g. electrical resistance of wires.

A *heat reservoir* has a uniform temperature throughout and this temperature is not altered by the transfer of any quantity of heat to or from the reservoir.

3.3 Work → Heat

It has been found by experiment that we may convert work into an equivalent amount of heat continuously and indefinitely. That is, work may be converted continuously and indefinitely into heat with 100% efficiency.

To convince ourselves of the validity of this statement, we have only to think of Joule's experiments to determine the mechanical equivalent of heat. In all these experiments, work of some kind, mechanical or electrical, is done which gives rise to heat. The amount of heat produced may be calculated by measuring the temperature rise and heat capacity of the system. The conversion of work into heat is continuous and we may produce heat indefinitely by doing work.

3.4 Heat → Work

The study of the conversion of heat into work lead Kelvin and Planck to the second law of thermodynamics.

Consider the reversible isothermal expansion of an *ideal* gas — ideal because the internal energy is not then a function of volume and, hence, will not alter when work is done isothermally. Equation (2.57) then holds, i.e.

$$\text{đ}q = \text{đ}w$$

Therefore the heat absorbed from the surroundings or some other heat reservoir may be converted into work with 100% efficiency. However, since there is a change of state of the gas, there will come a time when the pressure inside the system is equal to that of the surroundings and the process of expansion will stop.

It is therefore apparent that, if we wish to convert heat into work continuously and indefinitely, a process must be used which will, once a certain amount of work has been done, return the system to its original state ready to commence the whole process again. In other words, we require some type of cyclic process.

Consider an ideal heat engine E operating in a cycle. In one cycle, let the engine extract heat q_1 from a hot reservoir, do work w and reject heat $-q_2$ to a cold reservoir (see Figure 3.3). By 'hot' and

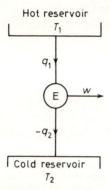

Figure 3.3. Engine E *operating between reservoirs at temperatures* T_1 *and* T_2

'cold', we mean that the temperature T_1 of the hot reservoir is greater than the temperature T_2 of the cold reservoir,

$$T_1 > T_2$$

[Notice that we are still using the sign convention: *heat absorbed by* a system (working substance) is *positive*; *work done by* a system (working substance) is *positive*. (There is no real need to use a sign convention so long as we make certain that energy is conserved.)]

The efficiency of an engine is given by

$$\eta = \frac{\text{work done by engine}}{\text{heat absorbed by engine}} = \frac{w}{q_1} \tag{3.1}$$

Since the working substance is returned to its original state, there is no change in the internal energy of the system,

$$\Delta U = 0 \tag{3.2}$$

Hence, from the first law,

$$q_1 - q_2 = w \tag{3.3}$$

From Eqs (3.1) and (3.3), we see that the efficiency of the engine E is

$$\eta = \frac{q_1 - q_2}{q_1} \tag{3.4}$$

i.e.

$$\eta = 1 - \frac{q_2}{q_1} \tag{3.5}$$

Therefore, if the engine is to be 100% efficient, $-q_2$ must be zero. Although this is possible in principle, in practice it is impossible. No engine has ever been devised which will indefinitely and continuously absorb heat from a reservoir and do an equivalent amount of work (Figure 3.4). It is this failure to convert heat into work

Figure 3.4. Perpetual motion of the second kind

continuously and indefinitely with 100% efficiency which constitutes the second law.

3.5 Refrigerator

If we could *operate the engine E of Figure 3.3 in reverse* it would *act as a refrigerator* (see Figure 3.5).

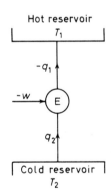

Figure 3.5. Refrigerator

Work $-w$ is done, the engine absorbing heat q_2 from the cold reservoir and rejecting heat $-q_1$ to the hot reservoir. Over a cycle, $\Delta U = 0$, so that

$$\text{net heat absorbed in a cycle} = \text{work done in a cycle}$$

$$q_2 - q_1 = -w$$

Hence,
$$-q_1 = -w - q_2 \tag{3.6}$$

i.e.
$$q_1 = q_2 + w \tag{3.7}$$

More heat is rejected by the engine to the hot reservoir than is absorbed from the cold reservoir.

3.6   Statements of the second law

THE KELVIN–PLANCK STATEMENT

It is impossible to construct an engine operating in a cycle that will produce no effect other than the extraction of heat from a reservoir and the performance of an equivalent amount of work.

If we could construct such an engine, then it would act simultaneously as a refrigerator and a motor with 100% efficiency.

This statement may be illustrated by the fact that it is impossible to run a ship by extracting heat from the sea, which obviously has an enormous amount of internal energy.

The continuous extraction of heat from our surroundings to do work is known as *perpetual motion of the second kind.* This type of perpetual motion, where heat is continuously extracted from the reservoir and an equivalent amount of work done, is shown in Figure 3.4. The *second law* therefore *denies* the possibility of *perpetual motion of the second kind.* It is perfectly possible to construct an engine to do the opposite, i.e. do nothing but convert work into heat completely.

Note that a ship could run on heat from the sea provided that a cold reservoir, i.e. a reservoir whose temperature was lower than that of the sea, were available to which heat could be rejected (see Figure 3.3).

THE CLAUSIUS STATEMENT

It is impossible for an engine operating in a cycle, unaided by an external agency, to transfer heat from one body to another at a higher temperature.

In other words, *heat will not pass spontaneously from one body to another at a higher temperature.*

If it is required to transfer heat from one body to another at a higher temperature then work must be done, as in the refrigerator.

*3.7   Proof of the equivalence of the Kelvin–Planck and Clausius statements

The Kelvin-Planck and the Clausius statements do not appear, at first sight, to be equivalent. We shall prove that they are so by showing that a contradiction of either leads to a violation of the other.

CONTRADICTION OF CLAUSIUS STATEMENT→ VIOLATION OF
KELVIN–PLANCK STATEMENT

Consider that an ideal engine E_1 operates in a cycle and, without any external help, transfers heat q_2 from a cold reservoir to a hot reservoir. That is, engine E_1 contradicts the Clausius statement. Let another ideal engine E_2 absorb heat q_1 from the same hot reservoir, do an amount of work w and reject heat $-q_2$ to the same cold reservoir (see Figure 3.6). The engine E_2 does not violate the Kelvin–Planck statement. If we combine the engines E_1 and E_2 to

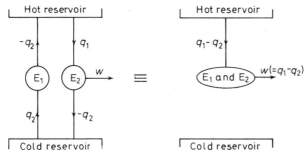

Figure 3.6. Engines E_1 and E_2 together violate the Kelvin–Planck statement

make a single engine, as in Figure 3.6, we find that this engine absorbs heat $q_1 - q_2$, from the hot reservoir but does not reject any to the cold reservoir. The heat absorbed is converted into an equivalent amount of work. This represents perpetual motion of the second kind and is impossible according to the Kelvin–Planck statement.

CONTRADICTION OF KELVIN–PLANCK STATEMENT→ VIOLATION OF
CLAUSIUS STATEMENT

Consider an ideal engine E_1, operating in a cycle, which absorbs heat q_1 from a hot reservoir and does an equivalent amount of work w, i.e. it does not reject any heat (see Figure 3.7). The engine E_1 contradicts the Kelvin–Planck statement. Let an ideal engine E_2 operate as a refrigerator such that, on doing work $-w$ ($= -q_1$), it absorbs heat q_2 from a cold reservoir and rejects heat $-q_1 - q_2$ to the hot reservoir (see Eq. 3.6). This engine does not violate the Clausius statement. If the two engines E_1 and E_2 are combined to make one engine (see Figure 3.7), we find that this machine transfers heat q_2 from the cold reservoir to the hot reservoir without any external aid, thus violating the Clausius statement.

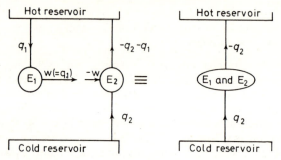

Figure 3.7. Engines E_1 *and* E_2 *together violate the Clausius statement*

We have shown that a contradiction of either statement leads to a violation of the other. Therefore, the two statements must be equivalent.

3.8   Reversible and irreversible processes and the second law

We have already stated that *a reversible process is a quasi-static (infinitely slow) process in which friction and all other similar means of energy dissipation are absent.* During a reversible process, the state of the system changes infinitely slowly, so that the system goes through a *continuous series of equilibrium states.* Reversible processes are hypothetical, ideal processes which cannot be achieved in practice. They are useful for comparison with real processes.

An alternative definition of reversibility will now be given: *A reversible process is one which is performed in such a way that if, at its conclusion, the direction of the process were reversed, then the system could be returned to its initial state leaving no changes whatsoever in the surroundings.*

Any process which does not satisfy this condition is said to be *irreversible.* In the following examples, we shall clarify the above definition of reversibility. It will also be seen that the two definitions are equivalent, and the connection between the second law and reversibility will become evident.

(1)   Consider the quasi-static adiabatic expansion of a gas in an insulated cylinder fitted with a piston. Let the piston be coupled to some external system on which work may be done as a result of the expansion.

The work done by the gas in the expansion is equal to the work done on the external system plus the work done against friction. The work required to return the gas to its *initial state*, by a quasi-static compression, is obviously more than that given to the external system, since friction opposes the motion both in expansion and compression. To obtain this additional amount of work without leaving changes in the surroundings, the heat produced by the friction must be converted completely into work. However, the second law states that the conversion of heat into work with 100% efficiency without producing changes elsewhere is impossible. Hence, at the conclusion of a quasi-static adiabatic expansion in the presence of friction, it is impossible to return the system and its surroundings to their initial states. Therefore this process is irreversible.

Note that the production of heat by friction *must* be an irreversible process. If it were reversible, then in the reverse process heat could be absorbed from a reservoir and an equivalent amount of work done, thus contradicting the second law. All processes in which friction or any other similar means of dissipating energy is present, e.g. viscosity, electrical resistance, inelastic deformation, magnetic hysteresis, are therefore irreversible processes.

(2) Consider example (1) above in the absence of friction. At the conclusion of the adiabatic expansion, all that is required to return the gas to its initial state (by a quasi-static process) is that the external system does the same amount of work on the gas as it received in the forward process. This is permissible. Therefore a quasi-static adiabatic expansion in the absence of friction is a reversible process, since the system and its surroundings may be returned to their initial states.

(3) Consider the isothermal expansion of a gas in a cylinder fitted with a frictionless piston. Let the gas, which has a temperature T, be brought into contact with a heat reservoir of higher temperature $T + \Delta T$ (Fig. 3.8). During the expansion, heat flows from the reservoir

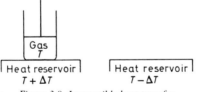

Figure 3.8. Irreversible heat transfer

to the gas. The system may be returned to its initial state by compression and the rejection of heat. This heat must be rejected to a different reservoir, the temperature of which is lower than that of the gas, by placing the cylinder on the second reservoir. Since the piston is frictionless, the work done during the expansion must be the same as that done during the compression. Although the system has been returned to its initial state, the surroundings, i.e. the two heat reservoirs, have not, since a quantity of heat has been transferred from the first to the second reservoir. This process is therefore irreversible.

For the isothermal expansion to be reversible, the system must reject the heat absorbed to the same reservoir. This can only occur when heat is transferred from a reservoir the temperature of which is only infinitesimally different from that of the system, i.e. the system and reservoir have essentially the same temperature (Figure 3.9).

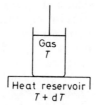

Figure 3.9. Reversible heat transfer

(4)   Consider the free expansion of an ideal gas. An insulated container is used with two compartments, one of which is evacuated, the other containing the gas (Figure 3.10a). If the partition is punctured, the gas will expand until it fills the entire container (Figure 3.10b). No work is done in the expansion, since there is no opposing force; $w = 0$. No heat transfer occurs, since the container is insulated; $q = 0$. Therefore, from the first law (Eq. 2.6), there is no internal energy change, $\Delta U = 0$.

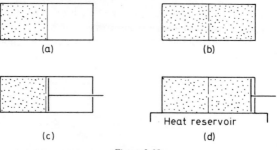

Figure 3.10.

If the process of free expansion were reversible, then it would be possible to proceed from the state where the gas occupies the whole container to that where it occupies half without producing changes in any external systems. Let us assume this to be true, so that from state (b) we may return to state (a) without leaving changes in the surroundings. If, at this stage, a piston were placed in the container (Figure 3.10c), and the gas allowed to expand (Figure 3.10d), then the work done by the gas would be equal to the decrease in the internal energy of the gas. The state (b) may be attained by the gas absorbing heat from a reservoir and hence increasing its internal energy. The cycle (b)→(c)→(d)→(b) is impossible according to the second law, since heat is absorbed and continuously converted into work with 100% efficiency. Therefore, the assumption that free expansion is a reversible process is incorrect.

Similarly, throttling and diffusion processes may also be shown to be irreversible.

Any process may be shown to be irreversible if the second law is contradicted on assuming the process to be reversible.

The fact that a process is irreversible does not necessarily mean that it cannot be reversed. The original state of a system can be restored at the conclusion of an irreversible process, but changes will be produced in the surroundings.

For a process to be reversible, the forces which the system and its surroundings exert on each other must be continuously balanced, and also any heat transfers must only occur with infinitesimal temperature differences.

Naturally occurring processes cannot be entirely free of friction or other forms of energy dissipation. Indeed, these processes may occur at a finite rate or involve the conduction of heat across a finite temperature difference. Since, during a reversible process, the system is always in a state of equilibrium, we come to the conclusion that *spontaneous processes are irreversible.*

A heat reservoir has a uniform temperature throughout and is capable of rejecting or absorbing any quantity of heat without altering its temperature. Therefore, when heat is absorbed or rejected by a reservoir, the changes occurring in the reservoir are the same as if the heat were transferred reversibly.

4

Entropy—
Classical Approach[†]

4.1 The Carnot cycle

The efficiency of an ideal steam engine is less than 20%. Why is the efficiency so low? Is it because (a) steam is used as the working substance; (b) the engine itself is inefficient; or (c) the inefficiency is due to something fundamental?

To answer these questions and others, Nicholas Leonard Sadi Carnot devised a set of very simple processes by which an ideal engine may operate in a cycle. The set of four reversible processes which constitute what is known as a *Carnot cycle* may be applied to any type of system.

During a Carnot cycle, an ideal engine operates between a hot reservoir, temperature T_1, and a cold reservoir, temperature T_2, the following four processes being performed in order:

Process 1 A *reversible isothermal process* in which the working substance *absorbs heat*, q_1, from the hot reservoir and the system does work.

Process 2 A *reversible adiabatic process* in which the *temperature* of the working substance *decreases* from T_1 to T_2 and the system does work.

Process 3 A *reversible isothermal process* in which the working substance *rejects heat*, $-q_2$, to the cold reservoir and work is done on the system.

Process 4 A *reversible adiabatic process* in which the working substance is *returned to its initial state*, the temperature of the system being altered from T_2 to T_1, and work is done on the system.

Over any cycle,
$$\Delta U = 0 \tag{4.1}$$

†This chapter may be omitted by those not wishing to approach entropy in a classical manner.

The work done, w, in one cycle = the net heat absorbed in one cycle

$$w = q_1 - q_2 \qquad (4.2)$$

An ideal heat engine operating in a Carnot cycle is known as a *Carnot engine*. In fact, *any reversible engine* operating between *only two* reservoirs must be a Carnot engine.

Figure 4.1 illustrates the result of one cycle of a Carnot engine C_R. The subscript R indicates that the four operations of a Carnot cycle are carried out reversibly. In one cycle, the engine absorbs heat q_1 from the hot reservoir, does work w and rejects heat $-q_2$ to the cold reservoir.

Since all four operations are reversible, the Carnot engine may be run in reverse as a refrigerator (see Figure 4.2). When operating

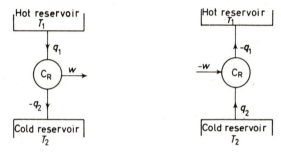

Figure 4.1. Carnot engine

Figure 4.2. Carnot engine acting as a refrigerator

in reverse, the Carnot engine pumps heat from the cold reservoir to the hot reservoir and may be referred to as a *heat pump*.

An ideal gas is the simplest working substance to which a Carnot cycle may be applied. Figure 4.3 shows the four operations of the Carnot cycle:

(1) The *reversible isothermal expansion* changes the state of the gas from A to B. Heat q_1 is absorbed at the temperature T_1 from the hot reservoir.

(2) The *reversible adiabatic expansion* changes the state of the gas from B to C, the temperature being altered from T_1 to T_2.

(3) The *reversible isothermal compression* changes the state of the gas from C to D. Heat $-q_2$ is rejected to the cold reservoir.

(4) The *reversible adiabatic compression* changes the state of the gas from D to A, the initial state, the temperature being altered from T_2 to T_1.

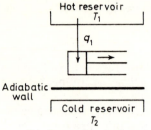

(1) Reversible isothermal expansion, system absorbs heat q_1

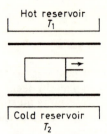

(2) Reversible adiabatic expansion from temperature T_1 to temperature T_2

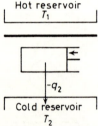

(3) Reversible isothermal compression, system rejects heat $-q_2$

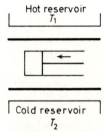

(4) Reversible adiabatic compression to initial state

Figure 4.3. The four reversible processes which, applied to an ideal gas in sequence, constitute a Carnot cycle

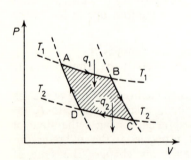

Figure 4.4. P–V diagram of a Carnot cycle applied to an ideal gas

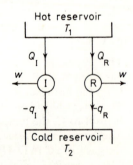

Figure 4.5. Irreversible and reversible engines operating between reservoirs at temperatures T_1 and T_2

These four operations may be shown on a $P-V$ diagram (see Figure 4.4). The work done in one cycle, w, is equal to the shaded area enclosed in the figure ABCD.

4.2 The Carnot theorem and corollaries

The Carnot theorem is as follows: *No engine operating between two heat reservoirs can be more efficient than a Carnot engine operating between the same two reservoirs.*

Consider that an engine I is more efficient than a Carnot engine R. That is,

$$\eta_1 > \eta_R$$

Let the two engines operate as follows in one cycle:

Engine I	Engine R
(1) absorbs heat Q_1 from hot reservoir	(1) absorbs heat Q_R from hot reservoir
(2) does work w	(2) does same amount of work w
(3) rejects heat $-q_1$ to the cold reservoir	(3) rejects heat $-q_R$ to the cold reservoir

Figure 4.5 illustrates these operations.

From the first law, since $\Delta U = 0$,

$$Q_1 - q_1 = w$$

$$q_1 = Q_1 - w$$

Similarly,

$$q_R = Q_R - w$$

If the engine I is more efficient than the reversible engine R, then it requires less heat to do the same work. Mathematically, if

$$\eta_1 > \eta_R$$

i.e.

$$\frac{w}{Q_1} > \frac{w}{Q_R}$$

then

$$Q_R > Q_1$$

Let the engine I drive the engine R in reverse (see Figure 4.6). This may be done without changing the heat and work quantities of R except for direction, since the engine R operates in a reversible cycle. The engine R then acts as a heat pump.

If we now combine engines I and R to make a single engine (see Figure 4.6), we find that heat $Q_R - Q_1$ is transferred from the cold to the hot reservoir without any external influence. This is impossible according to the second law. Therefore the assumption that the engine I is more efficient than engine R was wrong.

That is,
$$\eta_I \ngtr \eta_R$$

Hence,
$$\boxed{\eta_R \geqslant \eta_I} \tag{4.3}$$

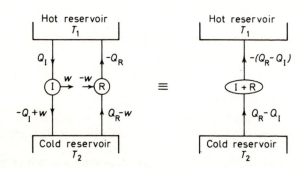

Figure 4.6. Proof of the Carnot theorem

Corollary 1   All Carnot engines working between the same two heat reservoirs are equally efficient.

Consider two reversible engines R_1 and R_2 working between the same two reservoirs. From the Carnot theorem, the engine R_1 cannot be more efficient than the engine R_2 and vice versa. Therefore the two engines must be equally efficient. That is,

$$\boxed{\eta_{R_1} = \eta_{R_2}} \tag{4.4}$$

Corollary 2   The efficiency of a Carnot engine is independent of the working substance.

Since all Carnot engines operating between the same two reservoirs have the same efficiency, the efficiency does not depend on construction details or on the working substance. Hence, the efficiency must depend only on the temperatures T_1 and T_2 of the two reservoirs. That is,

$$\boxed{\eta_{12} = f(T_1, T_2)} \tag{4.5}$$

where η_{12} is the efficiency of a Carnot engine operating between heat reservoirs at temperatures T_1 and T_2.

4.3 The thermodynamic temperature scale

Kelvin used the second law to define a thermodynamic temperature scale which is *independent* of the thermometric substance.

Consider a Carnot engine R_1 operating between the temperatures θ_1 and θ_2, these being temperatures on an arbitrary scale. Let the reversible engine R_1 absorb heat q_1 at θ_1, do work w_1 and reject heat $-q_2$ at θ_2 (see Figure 4.7).

$$\eta_{12} = \frac{q_1 - q_2}{q_1} = 1 - \frac{q_2}{q_1} = f(\theta_1, \theta_2) \tag{4.6}$$

Hence, q_2/q_1 must be a function of the temperatures θ_1 and θ_2,

$$\frac{q_2}{q_1} = F(\theta_1, \theta_2) \tag{4.7}$$

where $F(\theta_1, \theta_2) = 1 - f(\theta_1, \theta_2)$.

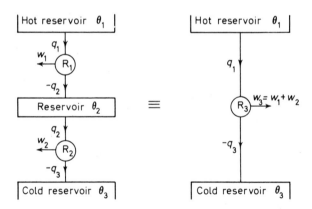

Figure 4.7. *Engines* R_1 *and* R_2 *are together equivalent to engine* R_3

Let a Carnot engine R_2 absorb heat q_2 at θ_2, do work w_2 and reject heat $-q_3$ at θ_3 (see Figure 4.7). Then, similarly,

$$\frac{q_3}{q_2} = F(\theta_2, \theta_3)$$

The two engines R_1 and R_2 are equivalent to a third engine R_3, working between θ_1 and θ_3, which absorbs heat q_1, rejects heat $-q_3$ and does work $w_3 = w_1 + w_2$, as in Figure 4.7.

$$\frac{q_3}{q_1} = F(\theta_1, \theta_3)$$

But

$$\frac{q_3}{q_1} \bigg/ \frac{q_3}{q_2} = \frac{q_2}{q_1}$$

Hence,

$$\frac{F(\theta_1, \theta_3)}{F(\theta_2, \theta_3)} = F(\theta_1, \theta_2) \tag{4.8}$$

The right-hand side of Eq. (4.8) does not contain θ_3, although the left-hand side does. Therefore, in order that the θ_3 terms cancel, the functions $F(\theta_1, \theta_3)$ and $F(\theta_2, \theta_3)$ must necessarily be of the form

$$F(\theta_1, \theta_3) = \frac{\psi(\theta_3)}{\psi(\theta_1)}$$

where the function $F(\theta_1, \theta_3)$ is the ratio of a function of θ_1 *only* to a function of θ_3 *only* and, similarly,

$$F(\theta_2, \theta_3) = \frac{\psi(\theta_3)}{\psi(\theta_2)}$$

Then

$$F(\theta_1, \theta_2) = \frac{\psi(\theta_3)}{\psi(\theta_1)} \cdot \frac{\psi(\theta_2)}{\psi(\theta_3)} = \frac{\psi(\theta_2)}{\psi(\theta_1)} \tag{4.9}$$

The ψ functions will be dependent on the temperature scale chosen. From Eqs (4.7) and (4.9),

$$\frac{q_2}{q_1} = \frac{\psi(\theta_2)}{\psi(\theta_1)} \tag{4.10}$$

Since the heat absorbed by the Carnot engine R_1 is greater than the heat rejected, i.e. $q_1 > q_2$,

$$\psi(\theta_1) > \psi(\theta_2)$$

Therefore the value of $\psi(\theta)$ increases with the temperature θ.

Kelvin used Eq. (4.10) to define a temperature scale such that

$$\frac{q_2}{q_1} = \frac{\psi(\theta_2)}{\psi(\theta_1)} = \frac{T_2}{T_1}$$

That is,

$$\boxed{\frac{q_2}{q_1} = \frac{T_2}{T_1}} \tag{4.11}$$

The ratio of two temperatures on the Kelvin scale is defined as the ratio of the heat rejected to the heat absorbed by a Carnot engine working between two reservoirs at those temperatures.

The thermodynamic temperature scale is based on the efficiency of the Carnot engine. Therefore it is independent of the thermometric substance, according to Corollary 2 of the Carnot theorem.

Notice that so far we have only defined the *ratio* of two temperatures on the thermodynamic (Kelvin) scale.

We shall later show that the thermodynamic temperature scale and the ideal–gas temperature scale are equivalent.

4.4 The efficiency of a Carnot engine

The efficiency of a Carnot engine working between a hot reservoir, temperature T_1, and a cold reservoir, temperature T_2, is, from Eqs (4.6) and (4.11),

$$\eta_{12} = 1 - \frac{q_2}{q_1} = 1 - \frac{T_2}{T_1}$$

$$\boxed{\eta_{12} = \frac{T_1 - T_2}{T_1}} \tag{4.12}$$

Since no heat engine can be more efficient than a Carnot engine working between the same two heat reservoirs, the efficiency given by Eq. (4.12) is the maximum possible and can *only* be attained by an ideal, reversible engine working between reservoirs at temperatures T_1 and T_2.

4.5 The absolute zero of temperature

It is seen from Eq. (4.12) that the efficiency of a Carnot engine increases with decrease in the temperature T_2 of the cold reservoir. *The absolute zero of temperature* is fixed on the thermodynamic scale as the temperature of the cold reservoir at which the efficiency of a Carnot engine becomes unity. That is, from Eq. (4.12), if $T_2 = 0$ then $\eta_{12} = 1$.

Since, from the principle of conservation of energy, no engine can be more than 100% efficient, the temperature of the cold reservoir cannot be negative. That is, from Eq. (4.12), since $\eta \not> 1$, T_2 cannot be negative.

In other words, negative temperatures cannot exist on the thermodynamic scale.

4.6   Temperature on the thermodynamic scale

Since absolute zero is fixed on the thermodynamic (Kelvin) scale, all that remains to be done is arbitrarily to fix the magnitude of a degree on this scale. This may be done by arbitrarily assigning a value for the temperature of some well-defined state. Such a state is the triple point of water, where water vapour, liquid water and ice are all in equilibrium with one another.

If either the temperature or pressure are altered even slightly from those at the triple point, this will cause at least one of the phases to disappear. The triple point of water is therefore a very well-defined state, and can be more accurately reproduced than the usual ice-point, which is pressure dependent. We shall arbitrarily assign the value of 273·16 K on the thermodynamic scale to be the temperature $T_{t.p.}$ at the triple point.

$$T_{t.p.} = 273 \cdot 16 \text{ K}$$

This now defines the magnitude of a Kelvin degree.

Any other temperature T on the thermodynamic scale may be defined by Eq. (4.11) by considering a Carnot engine operating between a reservoir at this temperature T and a reservoir whose temperature is 273·16 K.

The temperature 0°C is, on the Kelvin scale, 273·150 K.

4.7   Relationship between the thermodynamic temperature scale and the ideal gas temperature scale

Consider that the working substance of a Carnot engine be an ideal gas. The four operations of a Carnot cycle are shown in Figure 4.8, where T_1 and T_2 are the temperatures of the heat reservoirs on the ideal-gas temperature scale.

The gas expands isothermally from state A to state B at temperature T_1 and absorbs heat q_1 in the process. It then expands adiabatically from B to C, after which it is compressed isothermally from C to D at temperature T_2, rejecting heat $-q_2$. Finally, the gas is compressed adiabatically to its initial state A.

For a reversible, isothermal process, the heat absorbed by an ideal gas is given by Eq. (2.58). Applying this equation to the isothermal process A→B, the heat absorbed is

$$q_1 = nRT_1 \ln \frac{V_B}{V_A}$$

and, for the isothermal process C→D, the heat rejected is

$$-q_2 = nRT_2 \ln \frac{V_D}{V_C}$$

Combining these two equations,

$$\frac{q_2}{q_1} = \frac{T_2 \ln (V_C/V_D)}{T_1 \ln (V_B/V_A)} \qquad (4.13)$$

For the reversible adiabatic process B→C, Eq. (2.51) gives

$$T_1 V_B^{\gamma-1} = T_2 V_C^{\gamma-1}$$

Similarly, for the reversible adiabatic process D→A, we have

$$T_2 V_D^{\gamma-1} = T_1 V_A^{\gamma-1}$$

Hence,
$$\frac{V_B}{V_A} = \frac{V_C}{V_D}$$

Employing this result in Eq. (4.13), we have

$$\frac{q_2}{q_1} = \frac{T_2}{T_1} \qquad (4.14)$$

Equation (4.14) is of exactly the same form as Eq. (4.11), which was used to define the Kelvin temperature scale.

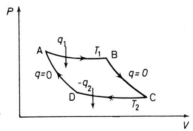

Figure 4.8. Carnot cycle

Combining Eqs (4.11) and (4.14),

$$\frac{T_2}{T_1} = \frac{T_2}{T_1}$$

By convention, the temperature of the triple point of water on the ideal gas scale is 273·16 K. Hence, temperatures as measured on the thermodynamic scale and on the ideal gas scale are numerically equal. That is,

$$T = T$$

We shall, as previously, use the symbol T to denote temperatures on both the Kelvin and the ideal-gas temperature scales.

Example   Find the maximum possible efficiency of a steam engine working between 373·15 K and 298 K.

Since no engine can be more efficient than a Carnot engine operating between the same two heat reservoirs, the maximum possible efficiency is given by Eq. (4.12), which is

$$\eta = \frac{T_1 - T_2}{T_1}$$

i.e. $$\eta = \frac{373·15 - 298}{373·15} = \frac{75·15}{373·15} = 0·201$$

The maximum possible efficiency is therefore 20·1 % but, since no practical engine can be as efficient as a Carnot engine, even this value cannot be attained.

*4.8   Replacement of a reversible process

Consider a reversible process, the path of which is represented by the solid line AB in Figure 4.9. The dotted lines through A and B

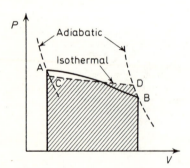

Figure 4.9. Replacement of a process by isothermal and adiabatic processes

represent reversible adiabatic processes and the line CD represents a reversible isothermal process such that

the area below ACDB = the area below the reversible path AB

In other words, we choose the isothermal process so that, in going from state A to state B, the work done via the path AB is equal to that done via path ACDB,

$$w_{AB} = w_{ACDB} \qquad (4.15)$$

From the first law,

$$Q_{AB} = U_B - U_A + w_{AB} \qquad (4.16)$$

$$Q_{ACDB} = U_B - U_A + w_{ACDB} \qquad (4.17)$$

where Q_{AB} and Q_{ACDB} are the quantities of heat transferred along the two paths.

Subtracting Eq. (4.16) from Eq. (4.17), and employing Eq. (4.15),

$$Q_{AB} = Q_{ABCD}$$

This means that the same amount of heat is absorbed along the two paths. Since there is no heat transfer in the two adiabatic processes AC and DB,

$$Q_{AB} = Q_{CD}$$

This result is very important, since it allows us to replace any reversible process in which the temperature may alter in any manner by a series of reversible isothermal and adiabatic processes such that the work done and heat transferred are the same. This is, in fact, true for any type of system, since generalised force and displacement could be written in place of P and V. In the next section, a *number* of adiabatic curves cutting the reversible cycle into a number of strips are chosen. Then isothermal curves are chosen such that a zig-zag curve is obtained which approximates to the original curve. They are also chosen so that, for each portion of the original curve, the same amounts of heat and work are transferred. The two isothermals between two adiabatic curves hence form part of a Carnot cycle.

4.9 Clausius' theorem

Consider a reversible cyclic process (Figure 4.10). The cycle may be broken up into a number of Carnot cycles so that for a particular isothermal curve, e.g. ab, of a given Carnot cycle, the heat transferred, q_5, is the same as that along the original curve cd of the reversible cycle. (Remember that there is no heat transferred in an adiabatic process.) The boundaries of the Carnot cycles form a zig-zag curve which approximates to the original cycle.

Let the first Carnot cycle absorb heat q_1 at a temperature T_1 and reject heat $-q_2$ at T_2. Then

$$\frac{q_1 - q_2}{q_1} = \frac{T_1 - T_2}{T_1}$$

$$\frac{q_1}{T_1} - \frac{q_2}{T_2} = 0 \tag{4.18}$$

Similar equations to Eq. (4.18) may be obtained for all the Carnot cycles. Summing all these equations,

$$\frac{q_1}{T_1} - \frac{q_2}{T_2} + \frac{q_3}{T_3} - \frac{q_4}{T_4} + \frac{q_5}{T_5} - \frac{q_6}{T_6} + \ldots = 0$$

$$\sum_i \frac{q_i}{T_i} = 0 \tag{4.19}$$

where the sign convention must be used, i.e. heat absorbed is positive.

If we increase the number of Carnot cycles used to describe our original cycle, the zigzag curve will approximate more closely to

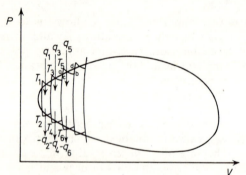

Figure 4.10. A cyclic process divided into a number of Carnot cycles

the original curve. In the limit, when infinitely narrow Carnot cycles are being used, i.e. infinitesimal reversible isothermal processes are used, the zigzag curve will be identical to the original curve. We must then replace the summation by an integration over the cycle, Eq. (4.19) becoming

$$\oint \frac{đq}{T} = 0 \tag{4.20}$$

where $đq$ is the heat transferred in an infinitesimal reversible isothermal change.

Equation (4.20) is known as *Clausius' theorem* and is true for any reversible cycle.

4.10 Clausius' inequality

Carnot's theorem states that the efficiency of an engine operating in an irreversible cycle between temperatures T_1 and T_2 is less than that of a reversible engine operating between the same two temperatures,

$$\eta_I < \eta_R \tag{4.21}$$

where η_I and η_R are the efficiencies of the irreversible and Carnot engines respectively.

Let the irreversible engine absorb heat q_1 at T_1 and reject heat $-q_2$ at T_2, so that

$$\eta_I = \frac{q_1 - q_2}{q_1}$$

The efficiency of the Carnot engine is given by Eq. (4.12):

$$\eta_R = \frac{T_1 - T_2}{T_1}$$

Substituting into (4.21), we have

$$\frac{q_1 - q_2}{q_1} < \frac{T_1 - T_2}{T_1}$$

$$\frac{q_1}{T_1} - \frac{q_2}{T_2} < 0 \tag{4.22}$$

Any cycle which is wholly or partly irreversible may be replaced by a number of infinitely narrow Carnot cycles. Hence,

$$\oint \frac{dq}{T} < 0 \tag{4.23}$$

The inequality (4.23) is known as *Clausius' inequality*. Combining (4.23) with Eq. (4.20),

$$\boxed{\oint \frac{dq}{T} \underset{R}{\overset{I}{\leqslant}} 0} \tag{4.24}$$

where I and R indicate that the inequality is to be used for irreversible cycles and the equality for reversible cycles, i.e. the integral over an irreversible cycle is less than zero and that over a reversible cycle is equal to zero.

4.11 Entropy

Let us examine Eq. (4.20) more closely:

$$\oint \frac{dq}{T} = 0 \text{ (for a reversible process)}$$

It has been pointed out in Section 1.15 (see Eq. 1.16) that a value of zero for a cyclic integral implies that the differential is that of a function of state.

We shall define a new function called *entropy*, denoted by S, such that changes in this function are given by

$$dS = \frac{dq}{T} \qquad \text{(reversible process) (4.25)}$$

From the discussion above and Eq. (4.20), it is clear that entropy is a function of state. It therefore has an exact differential and hence

$$S_2 - S_1 = \int_1^2 \frac{dq}{T} \qquad (4.26)$$

The heat must be transferred reversibly from state 1 to state 2.

The definition of this new thermodynamic function entropy involves the transfer of heat, which depends on the particular reversible path chosen. However, Eq. (4.20) tells us that the integral in Eq. (4.26) will be the same for all reversible paths between two states. Or, mathematically speaking, we know that the differential of heat is inexact but, from Eq. (4.20), we see that for a reversible process the differential dq/T is exact and therefore $1/T$ is the integrating factor for dq (see Eq. 1.34).

Equation (4.25) is said to be a mathematical formulation of the second law, since we have used the law to demonstrate the existence of the function of state we have called entropy.

Like internal energy and enthalpy, entropy is defined except for an arbitrary constant (see Eq. 4.26) which only defines a change in entropy).

This has been a classical approach to the concept of entropy. From Eq. (4.25), the entropy changes of a system may be calculated as shown in the following section. The next chapter will give another approach to entropy and in that chapter the significance of entropy will be made apparent. There is no reason why the student who fails to grasp the concept of entropy should not be able adequately to perform calculations. After all, the understanding of a more familiar quantity such as work is mainly through its definition, and to some extent our experience of the world.

4.12 Entropy calculations

It is important to remember that entropy is a function of state, and that therefore the entropy change between two given states is independent of the path between the states and whether or not the path is reversible. However, it is only for reversible processes that we can use Eq. (4.25) to calculate entropy changes. Hence, in order to calculate the entropy change for an irreversible process, it is necessary to find a reversible path between the two states involved.

We shall, in the following examples, distinguish between entropy changes of the system and of the surroundings. A system plus its surroundings is known as the universe, and this constitutes an isolated system.

Example Let us calculate the entropy change when 1 mole of water at 373·15 K (state 1) is converted into steam at 373·15 K (state 2), both states being at a pressure of 101 325 N m^{-2}. The molar latent heat of vaporisation of water, i.e. the enthalpy change per mole, Δh_{vap}, for the phase change at constant pressure, is 40·63 kJ mol^{-1}.

If we have a heat reservoir also at a temperature 373·15 K, the heat required for the phase change may be absorbed by the system reversibly. From Eq. (4.26), the entropy change for the system is

$$\Delta S_{system} = \int_1^2 \frac{\mathrm{d}q}{T}$$

Since the temperature of a system does not alter during a phase change, we can write

$$\Delta S_{system} = \frac{1}{T} \int_1^2 \mathrm{d}q$$

The heat absorbed by 1 mole of water in going from state 1 to state 2 is the molar latent heat of vaporisation. Therefore, we have

$$\Delta s_{\text{system}} = \frac{\Delta h_{\text{vap}}}{T} \qquad (4.27)$$

$$\Delta s_{\text{system}} = \frac{40\,630}{373 \cdot 15} = 108 \cdot 9 \text{ J K}^{-1} \text{ mol}^{-1}$$

The heat reservoir must have given up heat equal to $-\Delta h_{\text{vap}}$. Therefore the entropy change of the surroundings is

$$\Delta s_{\text{surroundings}} = \frac{-\Delta h_{\text{vap}}}{T}$$

$$= -108 \cdot 9 \text{ J K}^{-1} \text{ mol}^{-1}$$

The total entropy change, $\Delta s_{\text{total}} = \Delta s_{\text{system}} + \Delta s_{\text{surroundings}} = 0$
Further examples will be given in the following chapter.

4.13   Entropy change during an irreversible process

To find the entropy change for an irreversible process between two states, it is necessary to devise a reversible process for going from the same initial state to the same final state, so that Eq. (4.25) may be applied. The entropy change in going from one state to another is independent of the path, since entropy is a function of state.

Let a system change from state 1 to state 2 by either an irreversible process path A, or a reversible process path B, and return to state 1 by a reversible process path C (see Figure 4.11, where x and y are two appropriate functions of state).

From Clausius' equality, (4.24), we have for the cycle employing paths B and C

$$\int_{B\,1}^{2} \frac{\text{đ}q}{T} + \int_{C\,2}^{1} \frac{\text{đ}q}{T} \overset{\text{R}}{=} 0 \qquad (4.28)$$

where the subscript by the integral indicates the path, and for the cycle employing paths A and C

$$\int_{A\,1}^{2} \frac{\text{đ}q}{T} + \int_{C\,2}^{1} \frac{\text{đ}q}{T} \overset{\text{I}}{\underset{\text{R}}{\leqslant}} 0 \qquad (4.29)$$

Subtracting Eq. (4.28) from Eq. (4.29),

$$\int_{A\,1}^{2} \frac{\text{đ}q}{T} - \int_{B\,1}^{2} \frac{\text{đ}q}{T} \overset{\text{I}}{\underset{\text{R}}{\leqslant}} 0$$

so that

$$\int_{B_{J_1}}^{2} \frac{dq}{T} \geqslant \int_{A_{J_1}}^{2} \frac{dq}{T}$$

Since the path B is reversible, Eq. (4.25) may be employed, giving

$$\int_{1}^{2} dS \geqslant \int_{A_{J_1}}^{2} \frac{dq}{T}$$

or

$$dS \underset{R}{\geqslant} \frac{dq}{T} \qquad (4.30)$$

From the inequality, it is seen that for an irreversible process the entropy change is greater than dq/T, where dq is the heat transferred during the process.

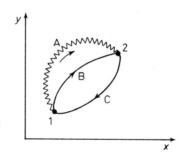

Figure 4.11. Reversible paths B and C and irreversible path A

Rewriting (4.30) as

$$T\,dS \geqslant dq \qquad (4.31)$$

it is interesting to compare this with Eq. (1.6), which states that $P\,dV \geqslant dw$.

For an isolated system, $dq = dw = 0$. From the first law, the system can only assume states for which the internal energy is the same. However, from (4.30), for an *isolated system*,

$$dS \underset{R}{\geqslant} 0$$

This equation states that the entropy of an isolated system either increases (for irreversible processes) or remains constant (for reversible processes). This is known as the *principle of entropy increase*. Hence, the entropy of an isolated system increases for a spontaneous process.

5
Entropy

5.1 Entropy changes in isolated systems

We have seen in Chapter 3 that the second law is a directional law which forbids the use of thermal energy in certain ways. As a result of this we found that *all spontaneous processes are irreversible*.

The information which may be obtained by applying the second law to a system plus its surroundings (universe) is (a) whether or not a process connecting two states of the universe may occur; (b) the direction of a process connecting two states of the universe; and (c) whether or not a process is reversible.

At the conclusion of an irreversible process, the system and the surroundings cannot be returned to their initial state and therefore the initial and final states of the universe must be in some way fundamentally different. On the other hand, at the conclusion of a reversible process, the universe may be returned to its initial state and therefore the initial and final states must in some way be equivalent.

Hence, on examination of the initial and final states, there must be some way of determining whether or not the universe can be returned to its initial state starting from the final state. Since we are interested in the initial and final states alone, we must be concerned with a change in some function of state. Therefore, the second law indicates the existence of a function of state which we shall call *entropy*. The entropy difference between the initial and final states of the universe will determine whether or not a process is reversible.

We have previously stated that for a reversible process, the initial and final states must be equivalent. Hence, if the entropies of the initial and final states of the universe are equal, i.e. the entropy difference is zero, then the process connecting these two states must be reversible. For an irreversible process, the entropy difference between the initial and final states of the universe will not be zero and shall be defined to be greater than zero. If the entropy difference

93

between some hypothetical initial and final state is negative, i.e. opposite to that for an irreversible process, then this will indicate that the process cannot take place in this direction. Therefore we have that a process altering the state of an isolated system may result in the *entropy change being zero*, in which case the *process is reversible*, or *greater than zero*, in which case the *process is irreversible*.

Using the symbol S for entropy, we have that, for an isolated system,

$$\Delta S \overset{\mathrm{I}}{\underset{\mathrm{R}}{\geqslant}} 0 \tag{5.1}$$

where the I and R remind us to use the inequality for irreversible, and the equality for reversible, processes. In the limit, considering infinitely small changes,

$$dS \overset{\mathrm{I}}{\underset{\mathrm{R}}{\geqslant}} 0 \tag{5.2}$$

Also, since we are dealing with an isolated system, the internal energy must be constant.

The second law really indicates the existence of *functions* of state such that, for the universe,

$d\psi = 0$ for a reversible process

$d\psi \neq 0$ for an irreversible process or an impossible process

where ψ is any of these functions of state.

The simplest of these functions is entropy S. (Others which we shall discuss later are the Gibbs and Helmholtz functions.) By saying that the entropy change of the universe for an irreversible process must be greater than zero, we are defining the sign of entropy.

Although entropy is a function of state and therefore an equation of the type

$$S = f(\text{independent variables})$$

must exist, it is not always explicitly known. In fact, only for ideal-gas systems are such equations easily attainable.

We shall justify later the following statement, which can be proved mathematically: *If, during an infinitesimal, reversible process, a system absorbs an amount of heat đq at a temperature T, then the entropy change of the system is*

$$dS = \frac{đq}{T}$$ (reversible) (5.3)

where T is the absolute temperature of the system.

Although heat is not a function of state, Eq. (5.3) states that $đq/T$ is an exact differential and therefore $1/T$ is the integrating factor for $đq$ (see Eq. 1.34). That is, for a reversible cycle,

$$\oint \frac{đq}{T} = 0$$ (5.4)

(Note that this is Clausius' theorem.)

Consider an irreversible process changing the state of a system from state A to state B. The entropy change is independent of the path of the process and whether or not the process is reversible since entropy is a function of state. However Eq. (5.3) applies to reversible processes only. The entropy change of a system between states A and B may be calculated by Eq. (5.3) *only* if a reversible path between the two states is devised.

$$S_B - S_A = \int_A^B \frac{đq}{T}$$ (reversible) (5.5)

It is always possible to find such a reversible path between two states of a system.

Equation (5.5) defines the function of state entropy except for an arbitrary constant. In this respect, entropy may be compared with internal energy and enthalpy.

5.2 Justification for $đq/T$ being an exact differential for a reversible process

This may be shown for an ideal gas system as follows. The first law, Eq. (2.5), states:

$$đq = dU + đw$$

For an ideal gas, $dU = C_V dT$, and, if the process is reversible, $dw = P dV$.

Hence,
$$dq = C_V dT + P dV$$

Dividing throughout by T, using $P = nRT/V$ and integrating from state 1 to state 2, we have

$$\int_1^2 \frac{dq}{T} = \int_1^2 \frac{C_V dT}{T} + \int_1^2 \frac{nR dV}{V}$$

$$\int_1^2 \frac{dq}{T} = C_V \ln \frac{T_2}{T_1} + nR\ln \frac{V_2}{V_1}$$

The right-hand side depends only on the initial and final states of the system. Therefore dq/T must be an exact differential. In a cyclic process, where the initial and final states are the same,

$$\oint \frac{dq}{T} = 0$$

JUSTIFICATION OF EQUATION (5.3) BY SHOWING THAT $dS \underset{R}{\overset{I}{\geqslant}} 0$

Let us find the entropy change when heat passes from a system A at temperature T_A to a system B at a lower temperature T_B (Figure 5.1). The two systems constitute the isolated system 1. The heat

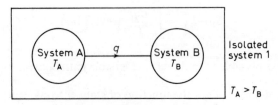

Figure 5.1. Irreversible heat transfer

flow is spontaneous and therefore irreversible. The entropy change of a system is independent of the process, being dependent only on the initial and final states. In order to make use of Eq. (5.3), we must find a reversible process having the same end-result as the spontaneous process.

Consider that the system A loses heat q to a heat reservoir infinitesimally lower in temperature than A and that the system B absorbs heat q from a reservoir infinitesimally higher in temperature

than *B* (Figure 5.2). The two systems plus the two reservoirs constitute the isolated system 2. As far as the systems A and B are concerned, the result of the reversible process is the same as that achieved
by the spontaneous process.

The entropy change of system A is

$$\Delta S_A = -\frac{q}{T_A}$$

(Remember that the heat absorbed is considered positive.)

The entropy change of system B is

$$\Delta S_B = \frac{q}{T_B}$$

Therefore, as a result of the *irreversible process*, the total entropy
change of the isolated system 1 in Figure 5.1 is

$$\Delta S_1 = \frac{q}{T_B} - \frac{q}{T_A}$$

Therefore $$\Delta S_1 > 0$$

since $T_A > T_B$.

This result is in agreement with the previous statement that the

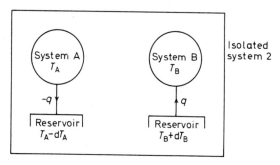

Figure 5.2. Reversible heat transfer

entropy change of an isolated system must be greater than zero
for an irreversible process.

For the process carried out reversibly, as in Figure 5.2, the entropy
change of the surroundings is

$$\Delta S_{surroundings} = \frac{q}{T_A} - \frac{q}{T_B}$$

The total entropy change of the isolated system 2 as a result of the *reversible process* is the sum of the entropy changes of the systems and surroundings,

$$\Delta S_2 = \underbrace{\frac{q}{T_B} - \frac{q}{T_A}}_{\text{systems}} + \underbrace{\frac{q}{T_A} - \frac{q}{T_B}}_{\text{surroundings}}$$

$$\Delta S_2 = 0$$

We have therefore found, using Eq. (5.3), that the entropy change of an isolated system is zero for a reversible process and greater than zero for an irreversible process, which is in agreement with our previous statements.

Equation (5.3) therefore satisfies all the requirements of the function of state entropy.

5.3   The second law

As those who have read Chapter 4 will realise, the second law not only indicates the existence of entropy but also of an absolute temperature scale. The absolute, or Kelvin, scale of temperature is independent of the thermometric properties of any system employed as a thermometer. It is based on the heat transferred during a Carnot cycle (to be discussed in Section 5.4).

In Chapter 4, the approach to entropy and absolute temperature (via Carnot cycles) may seem indirect, but in the present chapter we have merely shown the existence of entropy and have justified Eq. (5.3). A direct, but more abstract and mathematical approach, is that of Caratheodory. This only requires an alternative statement of the second law, this being that *in the neighbourhood of every state of a system there exist states which are inaccessible by means of a reversible adiabatic process.*

5.4   Temperature–entropy diagrams

From Eq. (5.3), the heat transferred in a reversible process is given by

$$Q = \int_1^2 T \, dS \qquad \text{(reversible process)} \qquad (5.6)$$

This integral may be interpreted graphically as the area under a reversible path on a temperature–entropy diagram (see Figure 5.3). This is similar to the representation of work on the $P-V$ diagram as the area under a reversible path (see Figure 5.4).

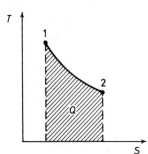

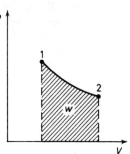

Figure 5.3. *T–S diagram,*
heat transferred = Q

Figure 5.4. *P–V diagram,*
work done = w

A reversible cyclic process on a $T-S$ diagram is shown in Figure 5.5. In the reversible process $1 \to 2$, the system absorbs heat equal to $\int_1^2 T\,dS$ (total shaded area). In the reversible process $2 \to 1$, the system rejects heat equal to $\int_2^1 T\,dS$ (see area cross-hatched only).

Therefore the area enclosed by the curve is equal to the heat absorbed in the reversible cycle.

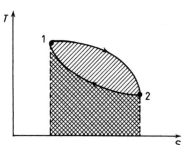

Figure 5.5. *Cyclic process on a T–S*
diagram

From the first law,

$$\oint đq = \oint dU + \oint đw$$

Since

$$\oint dU = 0$$

$$\oint đq = \oint đw \qquad (5.7)$$

Hence, the area enclosed by the curve is also equal to the work done over the cycle.

In the $T-S$ diagram (Figure 5.6), the *horizontal lines* $A \to B$ and $C \to D$ obviously represent *isothermal processes*.

During an adiabatic process, no heat enters or leaves a system; đ$q = 0$. Therefore, from Eq. (5.3), *for a reversible adiabatic process, there is no entropy change.* That is, reversible adiabatic processes are isentropic (constant entropy) processes. Hence, in Figure 5.6, the *vertical lines* B→C and D→A represent *reversible adiabatic processes.* The same cycle is also given on a *P–V* diagram (Figure 5.7) for a case when the system is an ideal gas. The set of four reversible processes in Figure 5.6, i.e. (1) reversible isothermal process, A→B, (2) reversible adiabatic process, B→C, (3) reversible isothermal process, C→D, (4) reversible adiabatic process, D→A, is known as a Carnot cycle. An engine operating in a Carnot cycle is known as a Carnot engine.

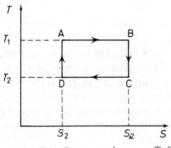

Figure 5.6. Carnot cycle on a T–S diagram

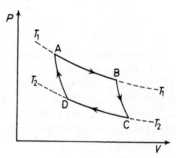

Figure 5.7. Carnot cycle on a P–V diagram

5.5   Efficiency of a Carnot engine

(The contents of this section are dealt with more fully in Chapter 4)
Carnot's theorem: No engine working between two heat reservoirs can be more efficient than a reversible engine operating between the same two reservoirs.

A reversible engine operating between *only two* heat reservoirs must be operating in a Carnot cycle. Since only reversible processes achieve the maximum work output and the minimum work input between two given states of a system, the maximum work is obtained from the system from state A to C and the minimum work is given to the system from C to A (Figures 5.6 and 5.7). Hence, Carnot's theorem must be true. Directly from the theorem follow two corollaries:
Corollary 1: All Carnot engines working between the same two reservoirs are equally efficient.

Corollary 2: The efficiency of a Carnot engine is independent of the working substance.

We shall now calculate the efficiency of a Carnot engine. From Figure 5.6, it is seen that (a) the heat absorbed in the reversible isothermal process A→B is $q_1 = T_1(S_2 - S_1)$; (b) the heat rejected in the reversible isothermal process C→D is $-q_2 = -T_2(S_2 - S_1)$. No heat transfer occurs in the adiabatic processes B→C and D→A. The work done in a cycle, w = the net heat transferred in the cycle

$$w = q_1 - q_2$$

$$= T_1(S_2 - S_1) - T_2(S_2 - S_1)$$

Therefore the efficiency of the Carnot engine is given by

$$\eta = \frac{w}{q_1}$$

$$\eta = \frac{(T_1 - T_2)(S_2 - S_1)}{T_1(S_2 - S_1)}$$

$$\boxed{\eta = \frac{T_1 - T_2}{T_1}} \tag{5.8}$$

This is the efficiency of a Carnot engine working between a heat reservoir at a temperature T_1 and another reservoir at a lower temperature T_2. Since no practical engine can be completely reversible, this is the maximum possible efficiency. Notice that if $T_2 = 0\,K$ then the efficiency of a Carnot engine is 100%. For a given temperature T_2, the efficiency is increased by increasing the temperature T_1 of the hot reservoir.

*5.6 Clausius' theorem and inequality

Consider a system undergoing a cyclic process which may be reversible or irreversible. The entropy change of the *system* is zero, since the system is returned to its original state:

$$\oint dS_{system} = 0 \tag{5.9}$$

We shall consider that during the process the system absorbs heat from a large number of heat reservoirs.

Consider an infinitesimal part of the cycle in which the system absorbs heat $\text{d}q$ from a heat reservoir at a temperature T. Then, since a reservoir can only absorb or emit heat reversibly (see Section 3.8), the resulting entropy change of the reservoir is

$$dS_{\text{reservoir}} = -\frac{\text{d}q}{T}$$

(minus, because the heat is rejected by the reservoir). The total entropy change of the universe for this infinitesimal change is

$$dS_{\text{universe}} = dS_{\text{system}} + dS_{\text{reservoir}}$$

$$dS_{\text{universe}} = dS_{\text{system}} + \left(-\frac{\text{d}q}{T}\right)$$

Summing up all the infinitesimal changes over the entire cycle, i.e. integrating over the cycle,

$$\oint dS_{\text{universe}} = \oint dS_{\text{system}} - \oint \frac{\text{d}q}{T}$$

Since for the system Eq. (5.9) holds, the entropy change of all the reservoirs involved in the cycle is, in this case, the entropy change of the universe. However, the entropy change of the universe must be greater than or equal to zero, so that

$$-\oint \frac{\text{d}q}{T}\Big|_{R}^{I} \geqslant 0$$

Therefore

$$\boxed{\oint \frac{\text{d}q}{T}\Big|_{R}^{I} \leqslant 0} \tag{5.10}$$

This result is a combination of Clausius' theorem and inequality and was, in fact, the statement given by Clausius of the second law.

5.7   Unavailable energy

Even under ideal conditions, not all the heat absorbed by an engine can be converted into work. Heat may therefore be regarded as consisting of available and unavailable parts for work.

Consider that a Carnot engine C absorbs heat q from a reservoir at a temperature T_1, does work w and rejects heat $-(q-w)$ to the coldest reservoir available for use, having a temperature T_0 (see Figure 5.8). Obviously, any energy contained in the coldest reservoir is completely unavailable for work. For a given amount of heat, the maximum work is obtained from a Carnot engine operating between the reservoir temperature T_1 and the *coldest* reservoir (Figure 5.9). The work done is

$$w_{max} = \text{heat absorbed} \times \text{efficiency}$$

$$= q\left(\frac{T_1 - T_0}{T_1}\right) = q\left(1 - \frac{T_0}{T_1}\right)$$

$$\boxed{w_{max} = q - T_0 \Delta S} \tag{5.11}$$

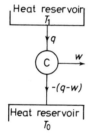

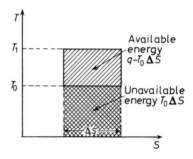

Figure 5.8. Engine C operating between two reservoirs, one of which has the lowest available temperature

Figure 5.9. Available and unavailable energy

The amount of heat $q - T_0 \Delta S$ which is used for work is referred to as available energy. The heat $T_0 \Delta S$ which is rejected to the coldest reservoir is referred to as unavailable energy.

Similarly, since any reversible process may be broken up into a number of Carnot cycles, it can be shown that the available energy is $\int dq - T_0 \Delta S$ (see Figure 5.10).

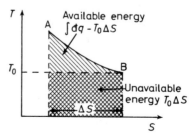

Figure 5.10. Available and unavailable energy

5.8   **The principle of dissipation of energy**

Consider that heat q is absorbed by a system B at a temperature T_B from a system A at a temperature T_A (Figure 5.11). Owing to the occurrence of this irreversible process, some of the heat q is not available for conversion to work.

The maximum work possible by converting heat q to work before conduction

$$\text{(available energy before conduction)} = q\left(1 - \frac{T_0}{T_A}\right)$$

where T_0 is the temperature of the coldest available reservoir.

The maximum work after conduction

$$\text{(available energy after conduction)} = q\left(1 - \frac{T_0}{T_B}\right)$$

$$\text{Loss of available energy} = q\left(1 - \frac{T_0}{T_A}\right) - q\left(1 - \frac{T_0}{T_B}\right)$$

$$= T_0\left(\frac{q}{T_B} - \frac{q}{T_A}\right) = T_0 \Delta S$$

where ΔS is the *entropy change of the universe* due to the irreversible conduction.

Now consider the heat q produced, by friction, at a temperature T, by a body sliding down a plane (Figure 5.12).

The entropy change of the universe is $\Delta S = q/T$.

Part of the heat q is available for work, and the maximum amount of work w would be produced by employing a Carnot engine and the coldest reservoir available, temperature T_0:

$$w = q\left(1 - \frac{T_0}{T}\right)$$

Hence, the waste of energy due to the irreversible process $= q(T_0/T)$ $= T_0 \Delta S$.

The above examples illustrate that *whenever an irreversible transfer or production of heat occurs, the effect on the universe is to convert energy available for work into a form unavailable for work. The amount of energy converted is the product of T_0 and the entropy change of the universe due to the irreversible process.*

This is known as the *principle of energy dissipation* and was developed by Kelvin. It must be understood that no *loss* of energy is implied: the principle of energy conservation must always hold. Energy merely becomes unavailable for the purpose of work.

Unfortunately, many scientists have considered that the principle of energy dissipation applied to *any* irreversible process and not just to those involving heat transfer or production. Hence, they have stated that the real meaning of the second law was embodied in the dissipation of energy principle. However, there are irreversible processes in which the energy of the system in the initial and final states is of exactly the same form.

For example, consider the free expansion of an ideal gas. An isolated container has two compartments, one being evacuated,

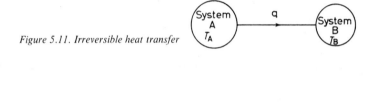

Figure 5.11. Irreversible heat transfer

Figure 5.12. Heat production by friction

the other containing an ideal gas. If the partition is punctured, the gas will rush into the vacuum and, hence, occupy the total volume.

No work is done in this irreversible process ($w = 0$), and, since no heat transfer occurs, $q = 0$. Hence, from the first law, there is no internal energy change: $U_{initial} = U_{final}$.

There is no dissipation of energy due to the irreversible process of free expansion. The energy of the system is not made in any way unavailable. Further examples are the mixing of two perfect gases and the dilution of a very dilute solution. These processes occur merely because there is an increase in entropy.

CLAUSIUS' STATEMENTS REGARDING THE UNIVERSE

Clausius states the first law as '*the energy of the universe is constant*' and the second law as '*the entropy of the universe tends to a maximum*'.

During irreversible heat transfers, energy becomes unavailable for work, and temperature differences tend to be equalised. Hence, Clausius predicted the thermal death of the universe.

5.9   The combination of the first and second laws

Consider a system in an equilibrium state at a temperature T. Suppose that owing to some process the system absorbs heat $đq$ and does work $đw$, and its internal energy increases by dU; then, from the first law,

$$đq = dU + đw \qquad (5.12)$$

From the second law, the total entropy change of the system and its surroundings must be greater than zero for an irreversible process or equal to zero for a reversible process:

$$dS + dS_s \overset{I}{\underset{R}{\geqslant}} 0 \qquad (5.13)$$

where S is the entropy of the system and S_s is the entropy of the surroundings. If we arrange that only reversible changes occur in the surroundings, then

$$dS_s = -\frac{đq}{T}$$

(minus because the heat $đq$ has been rejected by a reservoir).
Substituting for dS_s in Eq. (5.13),

$$dS - \frac{đq}{T} \overset{I}{\underset{R}{\geqslant}} 0$$

$$dS \overset{I}{\underset{R}{\geqslant}} \frac{đq}{T}$$

or

$$T\,dS \overset{I}{\underset{R}{\geqslant}} đq \qquad (5.14)$$

We may compare this equation with its equivalent in terms of work, $P\,dV \geqslant đw$. $T\,dS$ is the maximum heat that may be absorbed by a system under reversible conditions and $đq$ is the heat actually transferred. Employing Eq. (5.12),

$$T\,dS \geqslant dU + đw \qquad (5.15)$$

Rearranging, $\qquad \dbar w \leqslant T \, dS - dU$

This equation gives the upper limit to the amount of work which may be obtained from a system for an infinitesimal change.

If the only work permitted is that done reversibly by moving boundaries, then

$$\dbar w = P \, dV$$

and Eq. (5.15) becomes

$$\boxed{T \, dS = dU + P \, dV} \qquad (5.16)$$

Note that, strictly speaking, it is *wrong* to write

$$T \, dS \overset{\text{I}}{\underset{\text{R}}{\geqslant}} dU + P \, dV,$$

as is often done, since $P \, dV$ is the work done in an infinitesimal *reversible* expansion.

It is easily seen from Eq. (5.16) that entropy must be an *extensive* property, since this equation involves the sum of the extensive properties internal energy and volume.

Equation (5.16) may be derived directly from the first and second laws as follows. For an infinitesimal process,

$$\dbar q = dU + \dbar w \qquad (5.17)$$

For a reversible process,

$$\dbar w = P \, dV \qquad (5.18)$$

Hence, $\qquad \dbar q = dU + P \, dV \qquad (5.19)$

Now, since the second law requires the existence of entropy, we may regard the entropy change of a system for a reversible process, as given by

$$dS = \frac{\dbar q}{T} \qquad (5.20)$$

as a mathematical statement of the second law. Hence, on combining Eqs (5.19) and (5.20), we obtain Eq. (5.16).

Equations (5.18), (5.19) and (5.20) are true only for reversible processes. However, it is important to note that Eq. (5.16) involves only functions of state. That is, between two given equilibrium states S, T, U, V and P will alter only by the same amounts, irrespective of whether the process is reversible or not. Hence, Eq. (5.16) is not restricted to reversible processes alone.

A simple irreversible process which illustrates this is the free expansion of a gas. Consider an ideal gas in a compartment of an isolated vessel, the remainder of the vessel being evacuated. No work is done when the gas is allowed to expand freely and occupy the whole vessel: $w = 0$. Since the vessel is isolated, no heat enters or leaves the system: $q = 0$. Therefore, from the first law, the internal energy change of the gas is zero; $dU = 0$. However, $P\,dV$ is not zero. Also, since the process is spontaneous, the entropy change is not zero, which is in agreement with Eq. (5.16), from which it is seen the entropy change is $P\,dV/T$.

If the expansion were carried out reversibly, then, in order that the gas attain the same final state, it must, from the first law, absorb an amount of heat equal to $P\,dV$ from a reservoir at a temperature T. The entropy change would again be $P\,dV/T$.

5.10   Entropy changes

From Eq. (2.20), the heat absorbed during an infinitesimal isochoric process is

$$\text{đ}q = C_V\,dT$$

The entropy change during a reversible isochoric process is

$$dS = \frac{\text{đ}q}{T} = \frac{C_V}{T}\,dT \tag{5.21}$$

The entropy change of a system during a finite reversible process from state 1 to state 2 is

$$S_2 - S_1 = \int_1^2 \frac{C_V}{T}\,dT \tag{5.22}$$

Therefore, we may find $S_2 - S_1$ from a graph of C_V/T against T, or from a graph of C_V against $\ln T$.

Similarly, the entropy change during a constant pressure process is

$$\Delta S = \int_1^2 \frac{C_P}{T}\,dT \tag{5.23}$$

If the temperature dependence of the heat capacities is known, then the right-hand sides of Eqs (5.22) and (5.23) may be evaluated.

If, for example, C_P were temperature independent, then

$$\Delta S = C_P \int_{T_1}^{T_2} \frac{dT}{T}$$

$$\Delta S = C_P \ln \frac{T_2}{T_1} \qquad (5.24)$$

Or if

$$C_P = a + bT + cT^2 + \ldots$$

$$\Delta S = \int_{T_1}^{T_2} \left(\frac{a}{T} + b + cT + \ldots \right) dT$$

so that

$$\Delta S = a \ln \left(\frac{T_2}{T_1} \right) + b(T_2 - T_1) + \tfrac{1}{2}c(T_2^2 - T_1^2) + \ldots$$

$$(5.25)$$

Equations (5.22) and (5.23) are quite general and apply whether the system is a gas, a liquid or a solid.

5.11 Entropy changes of an ideal gas

Thermodynamically, an ideal gas has been defined as having (1) the equation of state $PV = nRT$; (2) no intermolecular forces, i.e.

$$\left(\frac{\partial U}{\partial V} \right)_T = 0. \qquad (5.26)$$

Equation (5.16) states that

$$T\,dS = dU + P\,dV$$

For an ideal gas, Eq. (2.39) states that

$$dU = C_V\,dT$$

Employing the equation of state of an ideal gas and integrating from state 1 to state 2, Eq. (5.16) gives

$$\int_1^2 dS = \int_1^2 \frac{C_V}{T} dT + \int_1^2 \frac{nR}{V} dV$$

$$\boxed{S_2 - S_1 = \int_1^2 \frac{C_V}{T} dT + nR \ln \frac{V_2}{V_1}} \qquad (5.27)$$

If C_V is temperature independent, then

$$S_2 - S_1 = C_V \ln \frac{T_2}{T_1} + nR \ln \frac{V_2}{V_1}$$ (5.28)

Or, substituting for volume using $V = nRT/P$ and also using Eq. (2.44), which states

$$C_P - C_V = nR$$

we have

$$S_2 - S_1 = C_P \ln \frac{T_2}{T_1} + nR \ln \frac{P_1}{P_2}$$ (5.29)

5.12　Probability

In order to obtain a more fundamental understanding of entropy and irreversible processes, some important ideas will be presented in the following sections.

Consider a ball in a box comprising two identical interconnecting compartments. After shaking the box well, there would be an equal probability of finding the ball in either compartment.

Probability is defined as *the ratio of the number of ways an event may occur to the total number of ways all possible events may occur.*

There are, in this case, two possibilities – the ball may be in one compartment or the other. Therefore, the probability of finding the ball in a given compartment is $\frac{1}{2}$. If there are two balls in the box, then, after shaking, the probability of finding a given ball in a particular compartment is $\frac{1}{2}$, the other ball being in the same compartment or the other. Therefore, the probability of finding both balls in a particular compartment is $\frac{1}{2} \times \frac{1}{2}$ or $(\frac{1}{2})^2$, i.e. the probability of two events happening is the product of the individual probability of each event. Following the same reasoning, we can see that if there are N balls in the box, the probability of finding all the balls in a particular compartment is $(\frac{1}{2})^N$. If $N = 20$, then the probability of finding all the balls in a particular compartment after shaking is approximately one in a million. If $N = 40$, then the probability of finding all the balls in a given compartment is one in 10^{12}.

Figure 5.13 shows the various possible arrangements for 1, 2, 3 and 4 balls in the box. Each arrangement is known as a microstate. To specify a microstate of a system, the state must be given to the ultimate limit of detail. To define the *microstate* of a thermodynamic system, the positions and energies of all the molecules in the system are required. To specify a *macrostate* of a system, an observer may experimentally measure the properties required. We need only a small number of parameters such as pressure, volume, temperature, etc. For a given macrostate, there is a large number of corresponding possible microstates.

If the balls are indistinguishable, then the two microstates $\boxed{a\,|\,b}$ and $\boxed{b\,|\,a}$ cannot be told apart by an observer, and therefore correspond to one macrostate.

In the case where there are 2 balls in the box, the probability of having a particular arrangement is $\frac{1}{4}$ and the probability of having 1 ball in each compartment is $2 \times \frac{1}{4}$. In the case of 3 identical balls in

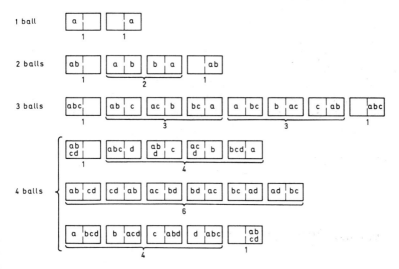

Figure 5.13. All the possible microstates when 1, 2, 3 and 4 balls are in a container with two identical interconnecting compartments

the box, there are 4 distinguishable macrostates of probabilities $\frac{1}{8}$, $\frac{3}{8}$, $\frac{3}{8}$, and $\frac{1}{8}$. If there are 4 identical balls in the box, there are 5 distinguishable states of probabilities $\frac{1}{16}$, $\frac{4}{16}$, $\frac{6}{16}$, $\frac{4}{16}$, $\frac{1}{16}$. The numerator of the probability gives the number of microstates corresponding to that particular macrostate.

It is easy to come to the conclusion that (a) as the number of balls is increased, the most probable macrostates are those with an equal number of balls in each compartment and (b) the two least probable macrostates are those where the balls are either all in one compartment or all in the other.

The results obtained by considering the balls in the box may be applied to molecules of a gas in a container divided into two identical compartments. The molecules of the gas will mix due to their thermal motions. If there are N molecules and they were all originally in one compartment, then the tendency would be to a state of maximum probability, that is, to the macrostate with the largest number of microstates. The larger N, the more probable is the state of even distribution and the less the probability of finding all the molecules in one compartment.

Due to the advent of quantum mechanics, we are now familiar with the concept of quantisation of energy. Let us consider a system consisting of 7 identical molecules which we shall consider to have been coloured by some mystical means (so that we can distinguish between them) red, orange, yellow, green, blue, indigo and violet. Let us consider that the total energy (i.e. internal energy) of the system is 7ε and that each of the molecules may have energy 0, ε, 2ε, 3ε, 4ε, 5ε, 6ε, or 7ε.

There are a number of ways of distributing the 7 molecules in the 8 energy levels to obtain a system of total energy 7ε. These distributions (macrostates) are given in Table 5.1. Consider the distribution 2. The molecule having the energy 7ε may be the red, while the rest have zero energy; or it may be the orange one, while the rest have zero energy. There are 7 possible ways of obtaining the distribution 2. That is, there are 7 microstates corresponding to the macrostate 2.

Table 5.1

Macrostate	Number of molecules having energy								Number of microstates corresponding to the macrostate W	Probability of macrostate
	0	ε	2ε	3ε	4ε	5ε	6ε	7ε		
1	–	7	–	–	–	–	–	–	1	0·000583
2	6	–	–	–	–	–	–	1	7	0·00408
3	5	1	–	–	–	–	1	–	42	0·02447
4	5	–	1	–	–	1	–	–	42	0·02447
5	5	–	–	1	1	–	–	–	42	0·02447
6	4	1	1	–	1	–	–	–	210	0·1224
7	4	1	–	2	–	–	–	–	105	0·06119
8	4	2	–	–	–	1	–	–	105	0·06119
9	4	–	2	1	–	–	–	–	105	0·06119
10	3	1	3	–	–	–	–	–	140	0·07973
11	3	2	1	1	–	–	–	–	420	0·2447
12	3	3	–	–	1	–	–	–	140	0·07973
13	2	3	2	–	–	–	–	–	210	0·1224
14	2	4	–	1	–	–	–	–	105	0·06119
15	1	5	1	–	–	–	–	–	42	0·02447

The number of microstates W corresponding to each macrostate may be found as above or, by calculation, from Eq. (5.30), which will be proved later in this section:

$$W = \frac{N!}{n_1!n_2!n_3!\ldots} \qquad (5.30)$$

where n_1 is the number of molecules having energy ε_1, n_2 is the number of molecules having energy ε_2, etc., and N is the total number of molecules in the system, i.e.

$$N = \Sigma_i n_i \qquad (5.31)$$

(Remember that $N! = N(N-1)(N-2)\ldots1$, e.g. $3! = 3 \times 2 \times 1$, and $0! = 1$.)

Let us now consider that the 7 molecules are in fact indistinguishable. Then, *although each particular microstate is as equally probable as any other, the macrostate which is most probable is 11, since this is the state with the maximum number of microstates* (420).

$$\text{The probability of the } i\text{th macrostate} = \frac{W_i}{\Sigma_i W_i} \qquad (5.32)$$

where $\Sigma_i W_i$ is the total number of microstates of the system.

Hence, the most probable state of the system is the macrostate with the maximum number of microstates, W_{max}. If the system had any other macrostate of lower probability and the system were left to itself, it would spontaneously change to the macrostate where W was a maximum (i.e. there would be a spontaneous change towards an equilibrium state).

The system we have considered consists of only 7 molecules. If we consider a larger system (e.g. 6×10^{23} molecules), then the probability of the macrostate which has the largest number of microstates is overwhelming.

The *thermodynamic probability* W of a given macrostate is defined as *the number of microstates corresponding to it*.

*PROOF OF EQUATION (5.30)

Let us consider a given macrostate. The 7 (coloured) molecules may be arranged in a number of ways, i.e. there are a number of microstates. Mathematically speaking, each arrangement is known as a permutation, the German word for which is *complexion*. Hence, sometimes the word complexion is used to mean microstate. Let us now place the 7 molecules one by one in the available positions, i.e. energy levels, of a given macrostate. The first position may be filled by using any one of the 7 molecules, i.e. there are 7 ways of filling the first position. The second position may be filled by choosing any one of the remaining 6 molecules, i.e. there are 7×6 ways of filling the first 2 positions. For each of the 42 possibilities for filling these 2 places, the third position may be filled in 5 ways, i.e. there are $7 \times 6 \times 5$ possibilities for the first 3 positions. For the 7 molecules, there are $7 \times 6 \times 5 \times 4 \times 3 \times 2 \times 1$ possible arrangements, i.e. 7! arrangements.

If there are N molecules in the system, there are $N!$ different arrangements (microstates). However, not all these molecules will be in different energy levels. Interchanging the position of any two molecules in the same energy level results in an identical microstate. This is equivalent to considering the molecules in the same level to be identical (i.e. having the same colour). Hence, if there are n_1 molecules in the energy level ε_1, there are $n_1!$ identical microstates, giving $N!/n_1!$ distinct arrangements.

Therefore, if there are

n_1 molecules in the energy level ε_1
n_2 molecules in the energy level ε_2

and finally

n_j molecules in the energy level ε_j

the number of microstates W for the given macrostate is given by

$$W = \frac{N!}{n_1! n_2! n_3! \ldots n_j!} \tag{5.30}$$

where $N = \Sigma_i n_i$.
Equation (5.30) may be written as

$$W = \frac{N!}{\prod_{i=1}^{j} n_i!} \tag{5.33}$$

Sometimes a given energy level may be realised in more than one way (i.e. more than one quantum state has the same energy). When this happens, the energy level is said to be *degenerate*. Let g_i be the degeneracy, or multiplicity, of the energy level ε_i. In other words, let there be g_i ways of obtaining the energy ε_i. If there is one particle in the ith level, then there are g_i possible distributions; if there are two particles, then there are g_i^2 possible distributions; if n_i particles, then $g_i^{n_i}$ possible distributions. Hence, the thermodynamic probability for a system of N particles is

$$W = N! \prod_i \frac{g_i^{n_i}}{n_i!} \tag{5.34}$$

5.13 Entropy and probability

Every system, if left to itself, i.e. without external influence, changes spontaneously (irreversibly) in such a way as to approach a state of equilibrium. Therefore, *all irreversible processes proceed in one direction only, namely towards an equilibrium state of the system.* We may regard the result of a spontaneous process as being to change the state of a system from a less probable to a more probable state. The *equilibrium state* of a system therefore *corresponds to the most probable macrostate* under a given set of conditions.

That there exists a relation between entropy and probability is suggested by the fact that, for a given set of conditions, both entropy and probability tend to a maximum during a spontaneous process for an isolated system. It was Boltzmann who suggested that entropy was a function of probability.

$$S = f(W) \tag{5.35}$$

where W is the thermodynamic probability of a state with entropy S.

Consider two systems A and B composed of the same material and under the same conditions, i.e. P and T. The total entropy is the sum of the entropies of the two systems, S_A and S_B,

$$S = S_A + S_B \tag{5.36}$$

The total thermodynamic probability of realizing a combination of the systems is the product of the individual probabilities of each system, W_A and W_B, since a given microstate of system A may be combined with each microstate of system B to give different microstates of the combined system:

$$W = W_A \cdot W_B \tag{5.37}$$

Hence, from Eqs (5.35), (5.36) and (5.37),

$$f(W_A \cdot W_B) = f(W_A) + f(W_B)$$

The only way this equation may be satisfied is by a logarithmic function. Therefore entropy is proportional to the logarithm of the thermodynamic probability,

$$S - S_0 = k \ln W \qquad (5.38)$$

where S_0 and k are constant (k, as we shall see later, being Boltzmann's constant). This is known as the Boltzmann–Planck equation.

Planck suggested putting $S_0 = 0$. Note that this corresponds to saying that for a state with a thermodynamic probability of unity, the entropy is zero. That is,

$$S = k \ln W \qquad (5.39)$$

A thermodynamic probability of unity means there is only one microstate of the system. That is, the state is a certainty.

If a system is *at equilibrium*, then the thermodynamic probability is a maximum, and, hence,

$$S = k \ln W_{max} \qquad (5.40)$$

However, when dealing with large systems, W_{max} is so large compared with the probability of any other particular macrostate that W_{max} may be taken to be almost equal to the total number of microstates of the system W_{tot},

$$W_{tot} = \Sigma_i W_i$$

Hence, for a system at equilibrium,

$$S = k \ln W_{tot} \qquad (5.41)$$

may sometimes be used as an approximation.

5.14   Probability and the second law

As previously explained, an irreversible process may be regarded as a process in which the system tends to a state of higher probability. If a system contains only a few molecules, it cannot be predicted that the system will pass from a given state to another state of higher probability. However, as we consider systems containing more and more molecules, our predictions will become increasingly accurate.

The second law of thermodynamics is therefore a statistical law and may be stated: *every system during a spontaneous process tends to a state of maximum probability.*

5.15 Isothermal expansion of an ideal gas

Consider 1 mole of an ideal gas (i.e. independent particles) which expands at constant temperature from volume V_1 to volume V_2. In 1 mole of gas, there are L molecules (L being known as Avogadro's constant). When the gas occupies volume V_2, the probability of

Figure 5.14. An ideal gas expands isothermally from volume V_1 to volume V_2

finding all the molecules within V_2 is unity. The probability of finding a particular molecule in the volume V_1 is V_1/V_2. The probability of finding *all* the L molecules within the volume V_1 is $(V_1/V_2)^L$.

The ratio of the probability of finding all the molecules within V_2 to that of finding all the molecules within V_1 is

$$1^L \left/ \left(\frac{V_1}{V_2} \right)^L \right.$$

which is $(V_2/V_1)^L$. Since thermodynamic probability is proportional to ordinary probability, this ratio must be equal to the ratio of the thermodynamic probability W_2 of the gas occupying volume V_2 to the thermodynamic probability W_1 of the gas occupying V_1 after expansion,

$$\frac{W_2}{W_1} = \left(\frac{V_2}{V_1} \right)^L$$

From Eq. (5.39),

$$S_2 - S_1 = k \ln W_2 - k \ln W_1 = k \ln \frac{W_2}{W_1}$$

$$S_2 - S_1 = Lk \ln \frac{V_2}{V_1} \tag{5.42}$$

Comparing this equation with Eq. (2.58), it is seen that

$$Lk = R$$

Therefore k in Eq. (5.39) is the Boltzmann constant, and, hence,

$$S_2 - S_1 = R \ln \frac{V_2}{V_1}$$

It is a simple matter to modify the derivation to deal with n moles of gas and arrive at

$$\boxed{S_2 - S_1 = nR \ln \frac{V_2}{V_1}}$$

*A STATISTICAL THERMODYNAMIC DERIVATION OF THE EQUATION OF STATE OF AN IDEAL GAS

Differentiating Eq. (5.42) with respect to V_2 and neglecting the symbol 2, we have

$$\left(\frac{\partial S}{\partial V}\right)_T = \frac{Lk}{V} \tag{5.43}$$

Equation (5.16), which combines the first and second laws, states

$$T \, dS = dU + P \, dV \tag{5.44}$$

If we assume that for an ideal gas at constant temperature the internal energy is constant, i.e. $dU = 0$, then Eq. (5.44) gives

$$\left(\frac{\partial S}{\partial V}\right)_T = \frac{P}{T}$$

From this equation and Eq. (5.43), we obtain $PV = LkT$. From a knowledge of ideal gases, it is known that k is the Boltzmann constant and that $Lk = R$ (as shown above). Hence,

$$PV = nRT$$

5.16 Entropy and disorder

A concept which may be found helpful is to regard the spontaneous tendency of a system to a state of maximum probability as the spontaneous tendency of a system to a state of maximum disorder for a given set of conditions. For example, consider a metal bar which is hotter at one end. The molecules at this end of the bar have higher energy than those at the other. This may be regarded as a more ordered state than that where the bar has reached a uniform temperature, when molecules of different energies are uniformly distributed. The mixing of two gases by diffusion may similarly be considered as a change from an ordered state (gases separated) to a state of disorder (gases mixed).

In the free expansion of a gas, the state of order so far as the vessel is concerned is that where the gas occupies one compartment of the vessel and that of disorder occurs after the expansion when the gas occupies the whole vessel.

Fusion and evaporation are both accompanied by entropy increases. In both cases, it may be considered that there is an increase in disorder. In fusion, the ordered crystal lattice structure of the solid is lost, and in vaporisation the ordered liquid structure is destroyed.

However, this concept of entropy and disorder being related may be carried too far. In general, it is *only true if* there are *no attractive forces* between molecules and there are *no external fields*, e.g. magnetic, electric or gravitational fields.

For example, consider a box containing a large number of identical plastic balls. If we place a small, strong bar magnet inside half the balls and shake the box, then we will find that the magnetic balls are grouped together. However, the state of maximum disorder corresponds to the magnetic and non-magnetic balls being completely mixed, which is a state of higher energy.

A similar experiment would be to consider a flask containing a polar liquid (e.g. water) and a non-polar liquid (e.g. petrol). In general, the two will not mix except at some high temperature, i.e. where the system has a higher internal energy so that the forces between the polar molecules may be broken.

Another example would be to consider a flask filled with water and a little sand. The sand will obviously sink to the bottom of the flask owing to gravity. However, maximum disorder corresponds to the sand and water being thoroughly mixed, which can only be achieved during violent shaking of the flask.

From these three examples, we see that the state of the system is subject to two opposing tendencies: (1) to maximum entropy (maximum disorder) and (2) to minimum internal energy (minimum disorder). For systems at relatively low temperatures, (2) predominates and the order increases, e.g. gases tend to liquids and liquids to solids at low temperatures. At relatively high temperatures, (1) predominates and the disorder increases. However, note that it is still true that a system always proceeds spontaneously to the state of maximum probability under the conditions present.

5.17   The meaning of entropy

Entropy, as we have seen, is an extensive function of state. For an infinitesimal reversible process, the entropy change is defined by

$$dS = \frac{\text{d}q}{T} \text{ (reversible)}$$

From this equation, it may be seen that the entropy of a system in a given state relative to the entropy of some arbitrarily selected reference state is found by considering a reversible process from the reference state to the actual state of the system. The entropy change is equal to the sum of the quotients of the heat transferred, $\text{d}q$, and the absolute temperature, T, at which the transfer occurs.

We have seen, in Section 5.1, that the entropy of an isolated system remains constant for a reversible process and increases for an irreversible process. That is, if the entropy change between state 1 and state 2 of an isolated system is greater than zero (i.e. $S_2 - S_1 > 0$), then the system will proceed spontaneously from state 1 to state 2. *The second law does not tell us how fast the spontaneous process occurs.* In order to know the rate, we must perform experiments.

Some spontaneous processes are extremely slow. For example, hydrogen and oxygen react spontaneously to give water;

$$H_2 + \tfrac{1}{2}O_2 \longrightarrow H_2O$$

The entropy of 1 mole of water is greater than the sum of the entropies of 1 mole of hydrogen and $\tfrac{1}{2}$ mole of oxygen gas under the same conditions. In the presence of a catalyst (sponge palladium), or if sparked, the reaction will proceed with explosive violence.

Entropy and thermodynamic probability are related by the Boltzmann–Planck equation,

$$S = k \ln W$$

Entropy is therefore a measure of the probability of a state. Alternatively, we have

$$W = e^{S/k}$$

We have also seen that entropy and the disorder of a system may be related. The spontaneous change of a system from one state to another of higher probability may be regarded as a change from a state of order to a state of greater disorder, in the absence of internal and external fields. It has been said that '*entropy is time's arrow*'.

5.18 Numerical entropy calculations

It is important to remember that entropy is a function of state and therefore the entropy change between two states is independent of the path between the states and whether the process is reversible or irreversible. It is only for reversible processes that Eq. (5.3) may be used to calculate entropy changes. To calculate the entropy change for irreversible processes, it is necessary to find a reversible path between the two states involved.

Example 1 Calculate the entropy change when 1 mole of ice at 273·1 K and 101 325 N m^{-2} pressure is converted into steam at 373·1 K and 101 325 N m^{-2}. The molar latent heat of fusion of ice is 6·00 kJ mol^{-1} at 273·1 K, the molar latent heat of vaporisation is 40·60 kJ mol^{-1} at 373·1 K and the molar heat capacity is 75·2 J K^{-1} mol^{-1} in the range 273·1–373·1 K.

For a reversible phase change,

$$\Delta S = \int_1^2 \frac{\mathrm{d}q}{T}$$

Since the temperature remains constant during a phase change, we may write

$$\Delta S = \frac{1}{T} \int_1^2 \mathrm{d}q$$

The heat absorbed by 1 mole of substance during the phase change is the molar latent heat, which is, of course, the molar enthalpy change Δh.

$$\Delta s = \frac{\Delta h}{T} \tag{5.45}$$

This equation is applicable to any phase change — fusion, vaporisation, sublimation and changes between crystalline forms — where T is the temperature at which the transition from one phase to another occurs.

The entropy change during a phase change may be found from Eq. (5.45). The entropy increase due to fusion

$$= \frac{6000}{273 \cdot 1} = 22 \cdot 0 \, \text{J K}^{-1} \, \text{mol}^{-1}$$

The entropy increase due to vaporisation

$$= \frac{40\,600}{373 \cdot 1} = 108 \cdot 8 \, \text{J K}^{-1} \, \text{mol}^{-1}$$

The increase in entropy due to heating 1 mole of water from 273·1 K to 373·1 K

$$= \int_{T_1}^{T_2} \frac{c_P}{T} \, dT = \int_{273 \cdot 1}^{373 \cdot 1} \frac{75 \cdot 2}{T} \, dT = 75 \cdot 2 \ln \frac{373 \cdot 1}{273 \cdot 1} \, \text{J K}^{-1} \, \text{mol}^{-1}$$

$$= 75 \cdot 2 \times 2 \cdot 303 \times \log_{10} \frac{373 \cdot 1}{273 \cdot 1} = 23 \cdot 5 \, \text{J K}^{-1} \, \text{mol}^{-1}$$

The total entropy change $= 22 \cdot 0 + 108 \cdot 8 + 23 \cdot 5 = 154 \cdot 3 \, \text{J K}^{-1} \, \text{mol}^{-1}$.

Example 2   The molar heat capacity of hydrogen is given by

$$c_P = 29 \cdot 05 - 0 \cdot 84 \times 10^{-3} \, T + 2 \cdot 10 \times 10^{-7} \, T^2 \, \text{J K}^{-1} \, \text{mol}^{-1}$$

Calculate the entropy change when 2 moles of hydrogen are heated from 273 K to 373 K at constant pressure.

The entropy change for 1 mole is

$$\Delta s_P = \int_{T_1}^{T_2} \frac{c_P}{T} \, dT = \int_{273}^{373} \left(\frac{29 \cdot 05}{T} - 0 \cdot 84 \times 10^{-3} + 2 \cdot 10 \times 10^{-7} \, T \right) dT$$

$$= 29 \cdot 05 \ln \frac{373}{273} - 0 \cdot 84 \times 10^{-3}(373 - 273) + \frac{2 \cdot 10 \times 10^{-7}}{2}(373^2 - 273^2)$$

$$= 8 \cdot 986 \, \text{J K}^{-1} \, \text{mol}^{-1}$$

The entropy change for 2 moles $= 2 \times 8 \cdot 986 = 17 \cdot 972 \, \text{J K}^{-1}$.

Example 3   Calculate the specific entropy change of water at a pressure of 101 325 N m^{-2} when heated from 200 K to 400 K, the latent heats of fusion and vaporisation being $3 \cdot 34 \times 10^5 \, \text{J kg}^{-1}$ and

22.6×10^5 J kg^{-1}, respectively, and the specific heat capacities being

$$c_P(\text{ice}) = 2.10 \times 10^3 \text{ J kg}^{-1} \text{ K}^{-1}$$

$$c_P(\text{water}) = 4.18 \times 10^3 \text{ J kg}^{-1} \text{ K}^{-1}$$

$$c_P(\text{steam}) = 2.10 \times 10^3 \text{ J kg}^{-1} \text{ K}^{-1}$$

The specific entropy change may be written as

$$\Delta s = \Delta s(\text{heating from 200 to 273 K}) + \Delta s(\text{melting})$$

$$+ \Delta s(\text{heating from 273 to 373 K}) + \Delta s(\text{vaporisation})$$

$$+ \Delta s(\text{heating from 373 to 400 K})$$

$$\Delta s_{200-273} = \int_{T_1}^{T_2} \frac{c_P}{T} \, dT = c_P \int_{T_1}^{T_2} \frac{dT}{T} = c_P \ln \frac{T_2}{T_1}$$

$$= 2.10 \times 10^3 \ln \frac{273}{200}$$

$$= 651 \text{ J kg}^{-1} \text{ K}^{-1}$$

Similarly,

$$\Delta s_{273-373} = 4.18 \times 10^3 \ln \frac{373}{273} = 1310 \text{ J kg}^{-1} \text{ K}^{-1}$$

and

$$\Delta s_{373-400} = 2.10 \times 10^3 \ln \frac{400}{373} = 146 \text{ J kg}^{-1} \text{ K}^{-1}$$

$$\Delta s(\text{melting}) = \frac{\Delta h(\text{melting})}{T_{\text{melting}}} = \frac{3.34 \times 10^5}{273} = 1230 \text{ J kg}^{-1} \text{ K}^{-1}$$

$$\Delta s(\text{vaporisation}) = \frac{\Delta h(\text{vaporisation})}{T_{\text{vaporisation}}} = \frac{22.6 \times 10^5}{373}$$

$$= 6060 \text{ J kg}^{-1} \text{ K}^{-1}$$

Therefore

$$\Delta s = 9397 \text{ J kg}^{-1} \text{ K}^{-1}$$

Problems

1. (a) The melting and boiling points of zinc are 692·5 K and 1180 K, and the latent heats of fusion and vaporisation are 7·520 kJ mol^{-1} and 115·8 kJ mol^{-1}, respectively. Calculate the entropy changes accompanying fusion and vaporisation.

Answer   10·87 J K^{-1} mol^{-1} and 96·5 J K^{-1} mol^{-1}

(b) At $101\,325\,\text{N m}^{-2}$ and in the temperature range $200\,\text{K}$ to $800\,\text{K}$, the molar heat capacity of zinc is given by

$$c_P = 21 \cdot 69 + 11 \cdot 15 \times 10^{-3}\,T\;\text{J K}^{-1}\,\text{mol}^{-1}$$

Calculate the molar entropy change when zinc is heated from $200\,\text{K}$ to $800\,\text{K}$.

Answer $35 \cdot 76\,\text{J K}^{-1}\,\text{mol}^{-1}$

2.   Calculate the molar entropy change when supercooled water at $263 \cdot 15\,\text{K}$ freezes at $101\,325\,\text{N m}^{-2}$, given the molar heat capacities of water and ice are $75 \cdot 3\,\text{J K}^{-1}\,\text{mol}^{-1}$ and $37 \cdot 6\,\text{J K}^{-1}\,\text{mol}^{-1}$, and the latent heat of fusion at $273 \cdot 15\,\text{K}$ is $6200\,\text{J mol}^{-1}$.

Answer $20 \cdot 65\,\text{J K}^{-1}\,\text{mol}^{-1}$

3.   One mole of an ideal gas at $300\,\text{K}$ expands isothermally from a pressure of $1\,000\,000\,\text{N m}^{-2}$ to $100\,000\,\text{N m}^{-2}$. If the expansion is (a) irreversible and frictionless against a pressure of $10\,000\,\text{N m}^{-2}$, (b) reversible, and (c) against zero pressure, calculate the heat absorbed, the internal energy change and the entropy change in each case.

Answer   (a) $225 \cdot 3\,\text{J mol}^{-1}$, 0, $17 \cdot 28\,\text{J K}^{-1}\,\text{mol}^{-1}$

(b) $5762\,\text{J mol}^{-1}$, 0, $17 \cdot 28\,\text{J K}^{-1}\,\text{mol}^{-1}$

(c) 0, 0, $17 \cdot 28\,\text{J K}^{-1}\,\text{mol}^{-1}$

4.   One mole of water at $373\,\text{K}$ is placed in a thermostated bath at $300\,\text{K}$. If the molar heat capacity of water is $75 \cdot 25\,\text{J K}^{-1}\,\text{mol}^{-1}$, calculate (a) the entropy change of the water and (b) the net entropy change.

Answer   (a) $-13 \cdot 7\,\text{J K}^{-1}\,\text{mol}^{-1}$, (b) $5 \cdot 18\,\text{J K}^{-1}\,\text{mol}^{-1}$

5.   One mole of water at $373\,\text{K}$ is mixed with one mole of water at $273\,\text{K}$ in a calorimeter. Calculate the entropy change if the molar heat capacity of water is $75 \cdot 25\,\text{J K}^{-1}\,\text{mol}^{-1}$.

Answer $1 \cdot 78\,\text{J K}^{-1}\,\text{mol}^{-1}$

6.   One mole of an ideal gas, $c_P = 29 \cdot 27\,\text{J K}^{-1}\,\text{mol}^{-1}$, is contained in a metal cylinder of heat capacity $83 \cdot 52\,\text{J K}^{-1}\,\text{mol}^{-1}$ at $300\,\text{K}$ and $2 \cdot 533\,125\,\text{MN m}^{-2}$. The gas expands adiabatically and reversibly to $101\,325\,\text{N m}^{-2}$. Calculate (a) the final temperature, (b) the work done, and (c) the entropy change.

Answer   (a) $236\,\text{K}$, (b) $6 \cdot 68\,\text{kJ}$, (c) 0

6
Ideal Gas Mixtures

6.1 Gibbs–Dalton law

We shall now consider how to evaluate the properties of a homogeneous mixture of ideal gases. We shall assume that the components of the system do not react chemically with one another. For a complete description of a mixture (multicomponent system), we require, in addition to the specification of two independent variables such as T and P, that the composition of the mixture be given in some manner.

The total mass of the mixture, m_{mix}, is the sum of the masses of the c components of the mixture:

$$m_{mix} = m_1 + m_2 + m_3 + \ldots + m_c = \sum_{i=1}^{c} m_i \qquad (6.1)$$

In a similar manner, the total number of moles of the mixture, n_{mix}, is the sum of the number of moles of each component:

$$n_{mix} = n_1 + n_2 + n_3 + \ldots + n_c = \sum_{i=1}^{c} n_i \qquad (6.2)$$

The mole fraction, x_i, of a component i is defined by

$$x_i = \frac{n_i}{n_{mix}} = \frac{n_i}{\Sigma_i n_i} \qquad (6.3)$$

Hence, from the definition and Eq. (6.2),

$$x_1 + x_2 + x_3 + \ldots + x_c = \sum_{i=1}^{c} x_i = 1 \qquad (6.4)$$

125

Dalton's law of partial pressures states that *the total pressure exerted by a mixture of ideal gases is equal to the sum of the partial pressures of each component.* The partial pressure of a gas is defined

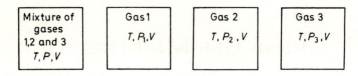

Figure 6.1. Illustration of partial pressure

as the pressure the pure gas would exert if it alone occupied the whole volume of the mixture at the same temperature.

That is, the total pressure is given by

$$P = P_1 + P_2 + P_3 + \ldots + P_c = \sum_{i=1}^{c} P_i \qquad (6.5)$$

where P_i is the partial pressure of gas i.

Now applying the equation of state for an ideal gas to each component, we have

$$P_i = n_i \frac{RT}{V} \qquad (6.6)$$

Hence, Eq. (6.5) gives

$$P = \Sigma_i n_i \cdot \frac{RT}{V} = \frac{RT}{V} \Sigma_i n_i \qquad (6.7)$$

From Eqs (6.6) and (6.7),

$$P_i = \frac{n_i}{\Sigma_i n_i} \cdot P$$

From Eq. (6.3),

$$\boxed{P_i = x_i P} \qquad (6.8)$$

The Amagat–Leduc Law states that *the volume of a mixture of ideal gases is equal to the sum of the partial volumes of the components of the mixture.* The partial volume of a gas is defined as the volume the pure gas would occupy if at the same temperature and pressure as that of the mixture.

That is,

$$V = V_1 + V_2 + V_3 + \ldots + V_c = \sum_{i=1}^{c} V_i \qquad (6.9)$$

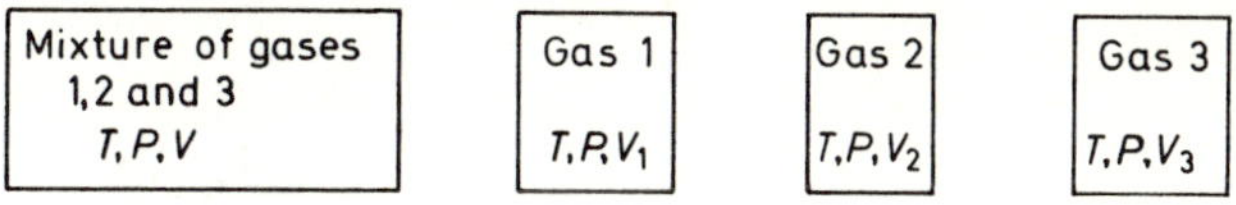

Mixture of gases 1,2 and 3 T, P, V	Gas 1 T, P, V_1	Gas 2 T, P, V_2	Gas 3 T, P, V_3

Figure 6.2. Illustration of partial volume

In a similar way to that employed in deriving Eq. (6.8), it can be shown that

$$\boxed{V_i = x_i V} \qquad (6.10)$$

In view of Eqs (6.1), (6.2) and (6.4) and the Dalton and Amagat–Leduc laws, it seems reasonable that a more general law exists which will enable us to evaluate the properties of ideal gas mixtures.

The Gibbs–Dalton Law states that, *in a mixture of ideal gases, each component of the mixture acts as if it alone occupied the volume of the system at the temperature of the mixture.*

This means that the presence of one component does not influence the behaviour of any of the other gases in the mixture. Microscopically, this implies that the intermolecular forces between the particles of the mixture are negligible. All the properties of mixtures of ideal gases may be obtained on the basis of the Gibbs–Dalton law.

The internal energy of the mixture is given by

$$U_{\text{mix}} = U_1 + U_2 + U_3 + \ldots U_c = \sum_{i=1}^{c} U_i \qquad (6.11)$$

where U_i is the internal energy of gas i.

Similarly, the enthalpy of the mixture is given by

$$H_{\text{mix}} = H_1 + H_2 + H_3 + \ldots + H_c = \sum_{i=1}^{c} H_i \qquad (6.12)$$

and the entropy of the mixture is given by

$$S_{\text{mix}} = S_1 + S_2 + S_3 + \ldots + S_c = \sum_{i=1}^{c} S_i \qquad (6.13)$$

The internal energy U_i of component i may be expressed as $n_i u_i$. Hence, Eq. (6.11) may be written as

$$n_{\text{mix}} u_{\text{mix}} = n_1 u_1 + n_2 u_2 + n_3 u_3 + \ldots n_c u_c = \sum_{i=1}^{c} n_i u_i \qquad (6.14)$$

where u_{mix} is the internal energy of the mixture per mole. Dividing through by n_{mix},

$$u_{\text{mix}} = x_1 u_1 + x_2 u_2 + x_3 u_3 + \ldots + x_c u_c = \sum_{i=1}^{c} x_i u_i \qquad (6.15)$$

Differentiating with respect to temperature at constant volume,

$$\left(\frac{\partial u_{\text{mix}}}{\partial T}\right)_V = x_1 \left(\frac{\partial u_1}{\partial T}\right)_V + x_2 \left(\frac{\partial u_2}{\partial T}\right)_V + \ldots + x_c \left(\frac{\partial u_c}{\partial T}\right)_V = \sum_{i=1}^{c} x_i \left(\frac{\partial u_i}{\partial T}\right)_V$$

That is, the molar heat capacity of the mixture is given by

$$c_{V,\text{mix}} = x_1 c_{V,1} + x_2 c_{V,2} + \ldots + x_c c_{V,c} = \sum_{i=1}^{c} x_i c_{V,i} \qquad (6.16)$$

Similarly, we may write Eq. (6.12) as

$$n_{\text{mix}} h_{\text{mix}} = n_1 h_1 + n_2 h_2 + \ldots + n_c h_c = \sum_{i=1}^{c} n_i h_i \qquad (6.17)$$

and a similar equation exists for entropy. From Eq. (6.17), we have

$$c_{P,\text{mix}} = x_1 c_{P,1} + x_2 c_{P,2} + \ldots + x_c c_{P,c} = \sum_{i=1}^{c} x_i c_{P,i} \qquad (6.18)$$

In general, we may say that, for any extensive property Z, the property for a mixture is given by

$$Z_{\text{mix}} = n_1 z_1 + n_2 z_2 + \ldots + n_c z_c = \sum_{i=1}^{c} n_i z_i \qquad (6.19)$$

where z_i is the property Z per mole of gas i when the gas occupies the entire volume of the mixture at the same temperature.

6.2　Entropy of a mixture

By considering the reversible separation of a gas mixture, it is a simple matter to show that the Gibbs–Dalton law holds when applied to entropy. Consider two identical cylinders, one of which can slide into the other without friction. The mixture is introduced when the cylinders are closed (Figure 6.3a). The wall AB, shown dotted, is permeable to gas 1 alone; similarly, CD is permeable to gas 2 only.

If an infinitesimal force is applied to the outer cylinder, so as to withdraw it infinitely slowly, the two gases will be separated reversibly. Let the partial pressures of the gases 1 and 2 be P_1 and P_2,

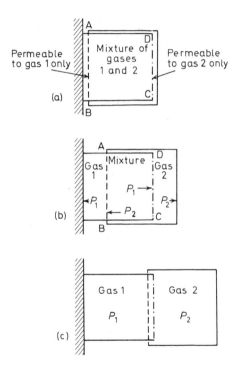

Figure 6.3. Reversible separation of a gaseous mixture

respectively. The pressure on the wall AB is solely due to gas 2, since gas 1 ignores this wall. Similarly, the pressure on the far wall of the outer cylinder is P_2. The pressure P_1 on the two vertical walls of the inner cylinder has no effect on the withdrawal of the outer cylinder.

Except for the initial infinitesimal force, the resultant force on the outer cylinder is zero. Since the initial force may be made as small as we like, the separation requires no work. No heat transfer occurs and therefore the internal energy and temperature of the system do not alter due to the process. The separation is reversible and hence the entropy of the system in the initial state must be equal to that in the final state,

$$S_{\text{initial}} = S_{\text{final}}$$

That is, the entropy of the mixture must be equal to the sum of the partial entropies of gases 1 and 2,

$$S_{\text{mix}} = S_1 + S_2$$

6.3   Entropy of mixing

Consider a container in which a number of ideal gases are separated from each other by partitions (Figure 6.4), the gases all being at the

Figure 6.4.

same temperature and pressure. The entropy of this system may be found by summing the entropies of all the gases. Employing Eq. (5.29) for each gas, we have that the entropy of the system is

$$S_{\text{sys}} = \Sigma_i \{ S_{0_i} + n_i c_{P,\,i} \ln T - n_i R \ln P \} \qquad (6.20)$$

where S_{0_i} is a constant.

If the partitions are now removed, the gases will mix spontaneously. The temperature of the system will remain constant since the internal energy of an ideal gas is independent of volume. The entropy of the mixture is given by

$$S_{\text{mix}} = \Sigma_i \{ S_{0_i} + n_i c_{P,\,i} \ln T - n_i R \ln P_i \} \qquad (6.21)$$

where P_i is the partial pressure of gas i.

The change in entropy due to mixing is

$$\Delta S_{\text{mixing}} = S_{\text{mix}} - S_{\text{sys}} = R\Sigma_i\{n_i \ln P - n_i \ln P_i\}$$

$$\Delta S_{\text{mixing}} = R\Sigma_i\, n_i \ln \frac{P}{P_i} \qquad (6.22)$$

Employing Eq. (6.8), we have that the entropy of mixing is

$$\Delta S_{\text{mixing}} = -R\Sigma_i\, n_i \ln x_i \qquad (6.23)$$

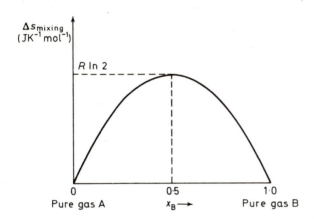

Figure 6.5. Graph of entropy of mixing against mole fraction

The entropy of mixing per mole of mixture is

$$\Delta s_{\text{mixing}} = -R\Sigma_i \frac{n_i}{n_{\text{mix}}} \ln x_i$$

That is,

$$\Delta s_{\text{mixing}} = -R\Sigma_i\, x_i \ln x_i \qquad (6.24)$$

From Eqs (6.23) and (6.10), we also have that

$$\Delta s_{\text{mixing}} = R\Sigma_i\, n_i \ln \frac{V}{V_i} \qquad (6.25)$$

It can be seen from Eq. (6.23) that ΔS_{mixing} is positive, since the mole fractions x_i must be less than one. This shows that the process of mixing is spontaneous.

Consider a two-component system, that is, the mixing of two ideal gases A and B, and let us compute ΔS_{mixing} from Eq. (6.24) for various compositions (see Fig. 6.5). It is seen that the maximum entropy of mixing is obtained when the gases are mixed in equal proportions. This, of course, corresponds to a system of maximum disorder, any other mixture being of greater order since it will be richer in one component.

Problem

1. Calculate the molar entropy change on the mixing of hydrogen and nitrogen gases, the mole fraction of hydrogen being 1/5.
Answer $15·07 \, J \, K^{-1} \, mol^{-1}$

7
The Third Law of Thermodynamics

7.1 Introduction

Integrating Eq. (5.3) from absolute zero to a temperature T, we obtain the entropy of the system at that temperature,

$$S = S_0 + \int_0^T \frac{\text{d}q}{T} \qquad (7.1)$$

where S_0 is the entropy of the system at 0 K. For many purposes, the value of S_0 is immaterial, since we are usually only interested in entropy changes. However, for phenomena involving chemical reactions or phase changes, S_0 appears explicitly in the relevant equations. Hence, we need to determine S_0 either by measurement or by calculation.

7.2 Nernst's heat theorem

As the temperature of a system in an equilibrium state is decreased to absolute zero, the entropy of the system tends to a constant value S_0, which is independent of the initial state of the system.

If S_0 were known, the entropy of a system in any state could be determined. The Nernst heat theorem may be stated mathematically as

$$\lim_{T \to 0} \Delta S = 0$$

That is, the change in the entropy of a system in an equilibrium state tends to zero as the absolute zero of temperature is approached (Figure 7.1).

Planck suggested that for every substance S_0 be put equal to zero:

$$S_0 = 0 \qquad (7.2)$$

(This is equivalent to normalising the entropy of all substances to zero at the absolute zero of temperature.)

Hence, Planck concluded that *the entropy of a pure, perfect, crystalline solid at absolute zero is zero*. This is known as *the third law of thermodynamics*.

The above statement is unduly restricting and a more general statement is: *the entropy of a system at absolute zero, in a state of thermodynamic equilibrium, is zero*. At temperatures approaching 0 K, the separation of the individual substances of a mixture is a spontaneous process. Therefore at very low temperatures a mixture is not in a state of equilibrium. The term *absolute entropy* is often employed to denote entropies which are referred to zero entropy at 0 K.

The historical development of the third law is rather remarkable for the number of controversies that arose. However, on close examination of the apparent exceptions to the law, it is found that,

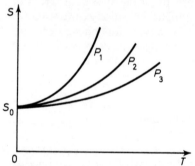

Figure 7.1. Entropy of a system as a function of temperature at various pressures

in all cases, the discrepancy may be attributed to the system being in a state of *apparent* equilibrium and undergoing an irreversible change. Some of these 'exceptions' will be discussed later.

In the first decade of the twentieth century, low-temperature studies had progressed sufficiently for it to become apparent that, as the absolute zero of temperature was approached, the heat capacities of real substances decreased rapidly to zero,

$$\lim_{T \to 0} C = 0$$

Since Eq. (7.1) may be written as

$$S - S_0 = \int_0^T \frac{C}{T} dT \qquad (7.3)$$

Nernst believed that his heat theorem could be derived directly from Eq. (7.3) but this is not so.

⌐ It should be noted that the third law is only obeyed by *real* substances and not, therefore, by an ideal gas or a van der Waals' gas. The reason for this is that the van der Waals equation of state, for example, is not applicable to real substances at very low temperatures.⌐

Following Einstein's ideas, Debye showed that at low temperatures (~ 15 K) the heat capacity of a solid may be expressed as

$$\boxed{C_P = aT^3} \qquad (7.4)$$

where a is a constant (see Chapter 16).

7.3 Determination of absolute entropies calorimetrically

Equation (7.3) states that, at constant pressure,

$$S - S_0 = \int_0^T \frac{C_P}{T} dT \qquad (7.5)$$

According to the third law, $S_0 = 0$ for a pure, perfect, crystalline substance:

$$\boxed{S = \int_0^T \frac{C_P}{T} dT} \qquad (7.6)$$

Therefore, if the heat capacity of a system is measured experimentally at various temperatures between T and the lowest available temperature T' (~ 10 K), the graph shown in Figure 7.2 may be plotted. Equation (7.6) may be written as

$$S = \int_0^{T'} \frac{C_P}{T} dT + \int_{T'}^T \frac{C_P}{T} dT \qquad (7.7)$$

The second integral in Eq. (7.7) may be determined by graphical integration. It is, of course, equal to the shaded area in Figure 7.2. To determine the first integral in Eq. (7.7), we make use of Debye's equation, which is a good approximation below about 15 K — that is,

$$C_P = aT^3$$

Hence,
$$\int_0^{T'} \frac{C_P}{T} dT = \int_0^{T'} aT^2 \, dT = \frac{1}{3} a(T')^3$$

$$= \frac{1}{3} C_P(T') \tag{7.8}$$

That is, the entropy below the temperature T' is approximately

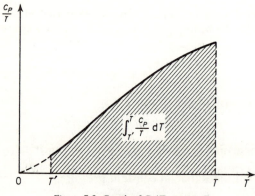

Figure 7.2. Graph of C_P/T against T

equal to 1/3 the heat capacity at the temperature T'. Eq. (7.7) may be written as

$$\boxed{S = \frac{1}{3} C_P(T') + \int_{T'}^{T} C_P \, d(\ln T)} \tag{7.9}$$

In the example given, there are no phase changes between the temperature T and 0 K. It is a simple matter to extend our ideas to include phase changes:

$$S = \int_0^{T'} C_P^{\text{solid}} \, d(\ln T) + \int_{T'}^{T_f} C_P^{\text{solid}} \, d(\ln T) + \frac{\Delta H_f}{T_f} + \int_{T_f}^{T_b} C_P^{\text{liquid}} \, d(\ln T)$$

$$+ \frac{\Delta H_V}{T_b} + \int_{T_b}^{T'} C_P^{\text{gas}} \, d(\ln T) \tag{7.10}$$

where T_f, T_b and T' are the melting and boiling points and the lowest available temperature, respectively,

$$C_P^{\text{solid}}, C_P^{\text{liquid}} \text{ and } C_P^{\text{gas}}$$

are the respective heat capacities, and ΔH_f and ΔH_v are the enthalpies of fusion and vaporisation (which are found from the molar latent heats).

The integrals in Eq. (7.10) could be evaluated by plotting a graph of C_P against $\ln T$ and determining the respective areas (see Figure 7.3).

Let us consider the determination of the molar entropy of oxygen gas at its boiling point, 90·13 K. There are three crystalline forms of oxygen.

Entropy change $(\text{J K}^{-1} \text{ mol}^{-1})$

Debye theory 0–14 K	2·26
Solid III, graphical integration 14–23·66 K	6·27
Transformation from solid III to solid II $= \dfrac{93\cdot70}{23\cdot66}$	3·96
Solid II, graphical integration 23·66–43·76 K	19·48
Transformation from solid II to solid I $= \dfrac{742\cdot0}{43\cdot76}$	16·97
Solid I, graphical integration 43·76–54·39 K	10·02
Fusion, $= \dfrac{444\cdot3}{54\cdot39}$	8·17
Liquid, graphical integration 54·39–90·13 K	27·01
Vaporisation, $= \dfrac{6804}{90\cdot13}$	75·50

Entropy of oxygen gas at 90·13 K $= 169\cdot14 \text{ J K}^{-1} \text{ mol}^{-1}$

With modern apparatus, it is possible to measure heat capacities down to about 1 K.

An experimental verification of the third law is possible by considering the entropy of a substance which may exist in more than one crystalline phase (allotrope). For simplicity, let us consider a substance having two crystalline phases, α and β. Below the transition temperature $T_{\alpha\beta}$, the β phase is metastable and therefore has the higher entropy, see Figure (7.4). The β phase may be obtained below $T_{\alpha\beta}$ by rapid cooling of the liquid.

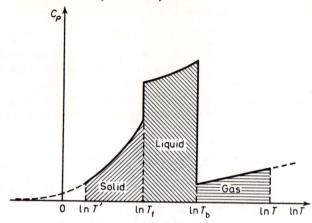

Figure 7.3. Graph of C_p against ln T

If, in accordance with the third law, the entropies of both phases are zero at absolute zero, then the entropy of phase β at a temperature greater than or equal to $T_{\alpha\beta}$ will be the same whether it is computed by studying the α phase or the β phase. The entropy may be found by measuring the molar enthalpy of the transformation of the α phase into the β phase and the molar heat capacities of the phases α and β as a function of temperature. By extrapolating (dotted lines in Figure 7.4) or calculation, the remaining entropies from the lowest available temperature down to 0 K may be found.

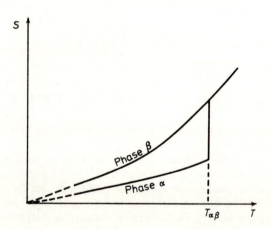

Figure 7.4. S–T diagram for a system which may exist in two phases

Tin may exist in two crystalline forms, white and grey, the latter being the stable form below room temperature. It is easy to keep white tin below the transition temperature 291 K.

At 0 K, the entropies of both grey and white tin are zero. Hence,

$$s_w = \int_0^{291} \frac{c_P^w}{T} \, dT \tag{7.11}$$

$$s_w = \int_0^{291} \frac{c_P^g}{T} \, dT + \frac{\Delta h}{291} \tag{7.12}$$

where s_w is the molar entropy of white tin at the transition temperature 291 K; c_P^w and c_P^g are the molar heat capacities of white and grey tin, both of which, of course, are functions of temperature, and Δh is the molar enthalpy for the transformation from grey to white tin.

By employing graphical integration to solve Eq. (7.11), s_w was found to be $52.0 \, \text{J K}^{-1} \, \text{mol}^{-1}$, and by employing graphical integration to solve Eq. (7.12), $51.8 \, \text{J K}^{-1} \, \text{mol}^{-1}$, which is in excellent agreement.

7.4 The Nernst heat theorem and the unattainability of absolute zero

An alternative statement of Nernst's heat theorem is: *It is impossible to reduce the temperature of any system to absolute zero in a finite number of operations.*

To attain low temperatures, the Joule–Kelvin effect is employed to produce liquid helium at a temperature of about 5 K. Rapid adiabatic evaporation reduces the temperature to about 1 K. Still lower temperatures are obtained by adiabatic demagnetisation.

Figure 7.5 shows how the entropy of a paramagnetic salt varies with temperature at zero magnetic field, $H = 0$, and magnetic field H. The curve for zero field must lie above that for field H, since in zero magnetic field there is no magnetic ordering. In a magnetic field, the magnetic dipoles are orientated in the same direction, hence greater order exists than in zero magnetic field and thus the system has lower entropy. According to the third law, both curves must pass through the origin.

The dotted curve is a hypothetical curve of the entropy the salt would be expected to have in zero magnetic field if Nernst's theorem were incorrect, i.e. if the two curves did not meet at 0 K. (In fact, since the ordering of dipoles is a spontaneous process at very low temperatures, this curve would be expected to meet the curve for field *H* at 0 K.) When the salt is isothermally magnetised, shown by line AB in Figure 7.5, the magnetic dipoles of the substance become orientated. If the salt is now adiabatically (and, hence, isentropically)

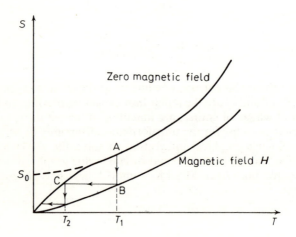

Figure 7.5. S–T diagram illustrating paramagnetic cooling

demagnetised, as shown by line BC, the temperature of the salt will fall. If Nernst's theorem were incorrect, i.e. at absolute zero the unmagnetised salt had an entropy S_0 different from that of the magnetised salt (see dotted line), then from the temperature T_1 it would be possible to attain absolute zero in one operation. If we now assume the alternative statement of Nernst's theorem above — that is, it is impossible to attain absolute zero — then a temperature T_2 must have been attained. Lower temperatures than T_2 may be achieved by further isothermal magnetisations and adiabatic demagnetisations, the temperature decrement becoming increasingly smaller. Hence, the two curves must meet, showing the equivalence of the two statements of Nernst's theorem. An infinite number or operations would be required to reduce the temperature to absolute zero. A similar discussion would be valid if the isobaric curves of Figure 7.1 did not meet at 0 K.

The temperature achieved by adiabatic demagnetisation is about 0·0014 K. Still lower temperatures of around 1×10^{-5} K have been achieved by applying the same technique to nuclear magnetic moments.

*7.5 Mathematical verification of Nernst's theorem using the unattainability of absolute zero

Consider a system which undergoes a reversible adiabatic process from state A to state B (Figure 7.6). This is an isentropic process — that is,

$$S_A = S_B$$

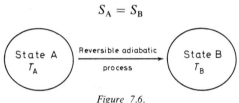

Figure 7.6.

Employing Eq. (7.3),

$$S_{A_0} + \int_0^{T_A} \frac{C_A}{T} \, dT = S_{B_0} + \int_0^{T_B} \frac{C_B}{T} \, dT \tag{7.13}$$

If absolute zero could be attained by this process, then $T_B = 0$, Eq. (7.13) becoming

$$\int_0^{T_A} \frac{C_A}{T} \, dT = S_{B_0} - S_{A_0}$$

Since the integral cannot be negative, this equation will hold if $S_{B_0} > S_{A_0}$. However, this conclusion must be incorrect since we have stated that it is impossible to attain 0 K. Hence,

$$S_{B_0} < S_{A_0}$$

Since the process is reversible, we may consider an adiabatic reversible process from state B to state A and show in a similar manner that $S_{A_0} < S_{B_0}$. Therefore, for both inequalities to be satisfied, satisfied, we must have

$$\boxed{S_{A_0} = S_{B_0}}$$

which is, of course, the Nernst theorem.

7.6   Entropy and degeneracy

From quantum mechanics, we know that as a system is cooled the constituent particles lose energy and fall into the lowest energy level. Hence, at absolute zero the entire system would be in its lowest energy level, which is known as the ground state. Now if the ground state is degenerate, i.e. if more than one quantum state can have this energy, then the constituent particles may be distributed in these quantum states to give more than one microstate at absolute zero. However, Eq. (5.39) states

$$S = k \ln W \tag{7.14}$$

and, hence, it is only when the system has just one microstate that the thermodynamic probability, W, is unity and, hence, $S = 0$.

Hence, the third law is equivalent to stating that the ground state of a system is non-degenerate.

Although it seems probable that this is true, so far no way has been found of proving it.

*7.7   Mixtures

If a system consists of a mixture of substances, then the entropy of the system will not be zero at 0 K, as there is entropy of mixing.

Consider a system in which we have a lattice consisting of v positions which are occupied by vx_A molecules of substance A and vx_B molecules of B, where x_A and x_B are the mole fractions of A and B, respectively. The number of possible microstates is given by Eq. (5.30);

$$W = \frac{v!}{(vx_A)!(vx_B)!} \tag{7.15}$$

In all practical cases, v will be a very large number. We may therefore employ Stirling's approximation:

$$N! \approx \left(\frac{N}{e}\right)^N$$

where N is a large integer.

Using this in Eq. (7.15) above, and remembering that $x_A + x_B = 1$,

$$W = \frac{\left(\dfrac{v}{e}\right)^v}{\left(\dfrac{vx_A}{e}\right)^{vx_A} \cdot \left(\dfrac{vx_B}{e}\right)^{vx_B}} = \left[\frac{1}{x_A{}^{x_A} \cdot x_B{}^{x_B}}\right]^v$$

$$\ln W = v[-x_A \ln x_A - x_B \ln x_B]$$

Employing Eq. (7.14), and multiplying and dividing by Avogadro's constant L,

$$S = \frac{vLk}{L}[-x_A \ln x_A - x_B \ln x_B]$$

But $v/L = n$, the total number of moles of A and B, and $Lk = R$, so that

$$S = nR[-x_A \ln x_A - x_B \ln x_B] \tag{7.16}$$

It is simple to extend the proof and consider a mixture of many substances, thus obtaining

$$S = -nR\Sigma_i x_i \ln x_i \tag{7.17}$$

It is seen from Eq. (7.17) that the entropy of a mixture is not zero at 0 K, even though the entropy of each individual pure component may be zero at 0 K.

7.8 'Exceptions' to the third law

Absolute entropies may be evaluated by graphical integration of heat capacity data as shown previously, or they may be determined from spectroscopic data with the aid of statistical mechanics (see Chapter 16).

The absolute entropies of some substances are given in Table 7.1. These are determined at 101 325 N m^{-2} (= 1 atm) pressure at the respective boiling points, with the exception of H_2O, D_2O, H_2 and D_2, determined at 298·15 K. There is excellent agreement between the calculated spectroscopic and measured calorimetric entropies for the first eight substances given. However, for the remaining compounds, the calorimetric entropy is less than the spectroscopic.

The reason for the discrepancies is that for these systems there is a residual entropy at absolute zero which is not taken into account in the calorimetrically measured entropies, owing to the systems being in non-equilibrium states. Thus these apparent exceptions to

Table 7.1

Substance	Boiling point (K)	Entropy as determined	
		spectroscopically	calorimetrically
A	87·3	129·0	128·7
O_2	90·13	170·0	170·0
N_2	77·4	153·2	153·6
HCl	188·2	173·2	172·6
Cl_2	238·6	215·4	215·6
CH_4	111·5	152·9	152·7
CO_2	194·7	198·8	198·9
NH_3	239·7	184·4	184·5
CO	83	160·1	155·5
NO	121·4	182·8	179·8
N_2O	184·6	202·7	198·0
CH_3D	99·7	165·0	153·5
H_2O	at 298·15	188·5	185·3
D_2O	at 298·15	194·7	191·8
H_2	at 298·15	130·5	124·1
D_2	at 298·15	144·7	141·6

the third law may all be explained on the basis that at sufficiently low temperatures the solids exist in disordered states.

(1) CO. The carbon monoxide molecule has a very small electric dipole. In the crystal, there is only a small difference in energy when the dipoles of two adjacent molecules are aligned in the same and in opposite directions, C—O C—O and O—C C—O. When, at low temperatures, the molecular rotations in the crystal cease, the dipoles will be randomly orientated and the molecules will possess insufficient energy to reorientate themselves. Hence, a state of disorder will exist in the crystal at 0 K. Since the orientation of the CO molecules is a spontaneous process at 0 K, the system is not in a state of equilibrium (although given a very long time at 0 K, an equilibrium state would be attained). A given molecule may be arranged in one of two directions within the crystal. Considering L molecules, where L is the Avogadro constant, i.e. considering

1 mole, there are 2^L microstates. The entropy at 0 K may be calculated by use of Eq. (7.14):

$$S = k \ln 2^L = R \ln 2$$
$$= 5{\cdot}77 \text{ J K}^{-1} \text{ mol}^{-1}$$

This value compares favourably with the difference of the spectroscopically and calorimetrically calculated entropies.

(2) N_2O. The same ideas are applicable to the linear nitrous oxide (N_2O) molecule which may have orientations NNO NNO and ONN NNO.

(3) NO. In the solid phase, nitric oxide, NO, exists as the dimer N_2O_2 which has two isomeric forms

$$\begin{array}{ccc} \text{N—O} & & \text{O—N} \\ \vdots \quad \vdots & \text{and} & \vdots \quad \vdots \\ \text{N—O} & & \text{N—O} \end{array}$$

with only slightly different energies. For L molecules of N_2O_2, there are 2^L microstates. For L molecules of NO, equivalent to $L/2$ molecules of N_2O_2, there are $2^{L/2}$ microstates. Hence, the entropy at 0 K is

$$S = k \ln 2^{L/2} = \frac{Lk}{2} \ln 2$$

$$= 2{\cdot}88 \text{ J K}^{-1} \text{ mol}^{-1}$$

This is in good agreement with the difference observed.

(4) CH_3D. There are four almost equivalent orientations in a crystal of CH_3D. The entropy at 0 K, which is $Lk \ln 4 = 11{\cdot}7 \text{ J K}^{-1} \text{ mol}^{-1}$, is in agreement with the observed difference.

(5) H_2O. The discrepancy here may be explained by assuming the presence of hydrogen bonding. In a water molecule, the H—O—H angle is 105° and the O—H bond length is 95 pm. It is assumed that, in ice, the water molecules are hydrogen-bonded together to form a crystalline lattice in which each oxygen atom is surrounded by four other oxygen atoms at the corners of a tetrahedron (Figure 7.7). The distance from the central oxygen atom to the corner of the tetrahedron, the O—O distance, is 276 pm.

Figure 7.7. A typical orientation of a water molecule

We assume that (a) only two hydrogen atoms are in the proximity of a single oxygen atom, i.e. directly bonded, and (b) only one hydrogen atom lies in each of the four O—O directions between the central oxygen atom and a tetrahedral oxygen atom, as in Figure 7.7.

Now, from assumption (b), each molecule has, along its four O—O directions, only two hydrogen atoms at about 95 pm. Hence, the chance of finding a hydrogen atom at about this position along a particular O—O direction from a given oxygen is $\frac{1}{2}$. The chance of finding a hydrogen atom at this position along two particular O—O directions from a given oxygen is $\frac{1}{4}$. However, it is seen from assumption (a) that a water molecule may be orientated in six ways relative

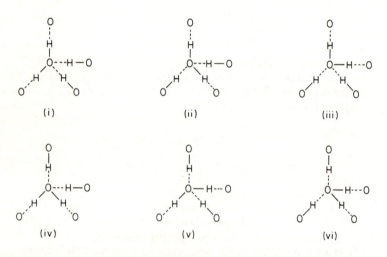

Figure 7.8. The six possible orientations of a water molecule

to its neighbouring oxygen atoms (Figure 7.8). The energy involved in the six orientations is approximately the same. Hence, the number of possibilities per molecule is $6 \times \frac{1}{4} = \frac{3}{2}$. The entropy per mole at 0 K is therefore $R \ln \frac{3}{2}$, which is in agreement with the observed difference for both water and heavy water, D_2O.

(6) H_2. Normally, hydrogen gas is composed of two molecular species since the two hydrogen atoms which together form a hydrogen molecule may have their nuclear spins either parallel or antiparallel. The two species are known as ortho- and parahydrogen, respectively. Since nuclear spins almost never change their orientation spontaneously, ortho- and parahydrogen are quite stable.

Hydrogen gas at room temperature is a mixture of ortho- and para-hydrogen in the ratio of 3:1. At 0 K there will be entropy of mixing, this accounting for the discrepancy observed. As the temperature drops, the composition of the equilibrium mixture changes, para-hydrogen being more stable at low temperatures. Normally, the equilibrium composition is not attained very rapidly, except in the presence of a catalyst (a paramagnetic substance).

SUPERCOOLED LIQUIDS AND GLASSES

At low temperatures, some supercooled liquids change spontaneously into the solid phase, others become glass-like in character. Glycerine provides an interesting example. Normally, it freezes at 291 K. However, it is possible to supercool glycerine down to about 180 K, at which temperature it becomes glass-like in character. The entropy of glassy glycerine at 0 K is about $23 \cdot 4 \, J \, K^{-1} \, mol^{-1}$ greater than that of crystalline glycerine.

The reason for this discrepancy is that, in a liquid, the molecules perform various motions, the degree of order being dependent on the temperature. When the supercooled liquid becomes glass-like, the viscosity increases, so that no further rearrangement to produce greater order is possible. Therefore, in the glassy phase, there is less order than in a solid crystalline phase. Any state in which the particles are disordered has a thermodynamic probability greater than one. Hence, the entropy at 0 K will not be zero.

Since the substance, as a glass, is in a supercooled state of disorder and therefore not in a state of thermodynamic equilibrium, the third law is not applicable. However, given time, it is possible for the molecules to achieve their true equilibrium positions.

ISOTOPE MIXTURES

Mixtures are apparent exceptions to the third law, owing to their possessing entropy of mixing at absolute zero. Similarly, if different isotopes of an element are present in a system, there will be a residual entropy at 0 K, e.g. 0^{16}, 0^{17} and 0^{18}, or Cl^{35} and Cl^{37}. It is only at temperatures well below 1 K that isotope separation becomes a spontaneous process. At these temperatures, the system is therefore no longer in thermodynamic equilibrium. The chemical properties of isotopes of a given element are very similar. Consequently, for most purposes it is usual to ignore this residual entropy unless there is a change in the isotope distribution as a result of the process studied.

7.9   Summary of the laws of thermodynamics

From the quantisation of energy principle, the internal energy of a system is given by

$$U = \Sigma_i\, n_i\, \varepsilon_i$$

where ε_i is the energy of the ith energy level
and n_i is the number of particles having energy ε_i.

The laws of thermodynamics may be summarised as follows, using the macroscopic and microscopic descriptions.

	Macroscopically (classically)	*Microscopically (statistically)*
First law	$dU = dq - dw$	$dU = \Sigma_i n_i d\varepsilon_i + \Sigma_i \varepsilon_i dn_i$
Second law	$dS = \dfrac{dq}{T}$	$dS = kd(\ln W)$
Third law	For a perfect crystal at absolute zero, $S = 0$	$S = k \ln W$ For a perfect crystal at absolute zero, $W = 1$—hence, $S = 0$

Problems

1.   Neglecting the effects of nuclear spins and considering non-equilibrium states, calculate the residual molar entropies at absolute zero of the following: (1) CO, (2) N_2O, (3) CH_3D, (4) CHD_3, (5) $CH_2{=}CHD$, (6) *trans* $CHD{=}CHD$.
Answer   (1) $R \ln 2$, (2) $R \ln 2$, (3) $R \ln 4$, (4) $R \ln 4$, (5) $R \ln 4$,
            (6) $R \ln 2$
2.   Discuss the significance of the statement that the absolute zero of temperature is unattainable.
3.   From the molar heat capacity data for silver given below, and Debye's equation, determine graphically the entropy of silver at 298 K.

T/K	c_p J K^{-1} mol^{-1}
15	0·669
30	4·77
50	11·64
70	16·31
90	19·11
150	22·94
250	24·69
300	25·46

Answer   42·68 J K^{-1} mol^{-1}

8
The Helmholtz and Gibbs Functions

8.1 Helmholtz and Gibbs functions

Using only the functions of state that have so far been defined and
the laws of thermodynamics, we could tackle any problem involving
a closed system. However, the resulting relationships would, in
many cases, be very complex. For mathematical simplicity, two
new functions of state are introduced: the Helmholtz function and
the Gibbs function.

The Helmholtz function, A, is defined by

$$A = U - TS \tag{8.1}$$

A is also known as the work function or Helmholtz free energy, and
is sometimes even referred to merely as free energy.

This definition Eq. (8.1), is not very informative as to the meaning
of the Helmholtz function. However, it can be seen that A is a function
of state, since it involves the difference and product of functions of
state, and also that it is an extensive property, since U and S are
extensive properties.

The Gibbs function, G, is defined by

$$G = H - TS \tag{8.2}$$

G is also known as Gibbs free energy or simply as free energy. In
order to avoid confusion, we shall refer to A as the Helmholtz
function and G as the Gibbs function.

Once again, little information can be obtained from the mathematical definition (Eq. 8.2) other than that G is an extensive function
of state since it involves the extensive properties H and S.

Both A and G may be regarded as energy functions. It is also seen from the definitions that A and G can only be known to within an arbitrary constant.

THE SIGNIFICANCE OF THE HELMHOLTZ FUNCTION

Consider the combination of the first and second laws (Eq. 5.15), which states

$$T \, dS \underset{R}{\overset{I}{\geqslant}} dU + \text{đ}w$$

Rearranging, we have

$$dU - T \, dS \underset{R}{\overset{I}{\leqslant}} -\text{đ}w \qquad (8.3)$$

Let us consider a system at *constant temperature*; then Eq. (8.3) may be written as

$$d(U - TS) \underset{R}{\overset{I}{\leqslant}} -\text{đ}w$$

From Eq. (8.1), we have that, at constant temperature,

$$\boxed{dA \underset{R}{\overset{I}{\leqslant}} -\text{đ}w} \qquad (8.4)$$

The work done *on* the system is greater than or equal to the increase in the Helmholtz function for an isothermal process, depending on whether the process is irreversible or reversible. Therefore the work done *on* the system during a reversible process is less than that done during an irreversible process between the same initial and final states. In other words, *the work done on a system in an isothermal process to bring about a given change of state is a minimum if the process is reversible.* It is seen from Eq. (8.4) that in an isothermal process the work done *by* the system is less than (irreversible) or equal to (reversible) the decrease in the Helmholtz function. That is, the work done by a system in an isothermal reversible process is greater than that done in an isothermal irreversible process between the same two states. Hence, for a given change of state, *the maximum work is done by a system in an isothermal process if that process is reversible.*

We are now in a position to understand the significance of the Helmholtz function. It is an extensive function of state, the decrease of which is equal to the maximum work obtainable from a system for an isothermal process between two states. Of course, this maximum work can *only* be obtained for a reversible process.

The definition, Eq. (8.2), may be written as

$$G = U + PV - TS \tag{8.5}$$

Employing Eq. (8.1),

$$G = A + PV \tag{8.6}$$

For an infinitesimal change in state of a system at *constant pressure*,

$$dG = dA + P \, dV \tag{8.7}$$

That is,

$$-dG = -dA - P \, dV$$

Applying the additional restriction of constant temperature, employing Eq. (8.4), we have that, at constant P and T,

$$-dG \geqslant \underset{R}{\overset{I}{đw}} - P \, dV$$

That is, the decrease in the Gibbs function for an isothermal–isobaric process is equal to the maximum work obtainable from the system, excluding the work done by expansion, for a given change of state. This maximum work would, of course, only be obtained during a reversible process. This maximum work excluding work done during volume changes is known as the maximum *useful work* — 'useful' since work done in expansion is done against the confining pressure (usually atmospheric). Hence, we may write

$$-dG = (đw_{\text{useful}})_{\text{max}} \tag{8.8}$$

8.2 Isothermal processes

From Eq. (8.1) it is seen that, for a finite isothermal change,

$$\Delta A = \Delta U - T \Delta S \tag{8.9}$$

Similarly, from Eq. (8.2), for a finite isothermal change,

$$\Delta G = \Delta H - T\Delta S \qquad (8.10)$$

In chemical processes the changes in both the internal energy, ΔU, and enthalpy, ΔH, usually far exceed the $T\Delta S$ term, except at high temperatures.

If a reaction is carried out reversibly at constant volume, the internal energy change ΔU may be converted *directly* into work, e.g. electrical work, or vice versa, depending on whether or not the reaction is exothermic. The same is true with respect to the enthalpy change when a reaction is carried out reversibly and isobarically. ΔU and ΔH may, of course, be obtained by measuring the heat of reaction when the reaction takes place irreversibly at constant volume and constant pressure, respectively. ΔU and ΔH may also be obtained by other means, e.g. using reversible cells (see Section 15.9).

Depending on its sign, the $T\Delta S$ term is equal to the heat absorbed or rejected to a heat reservoir when the reaction is carried out reversibly.

Example 1 To illustrate the above, consider the reaction

$$Zn + CuSO_4 \longrightarrow ZnSO_4 + Cu$$

If zinc is dropped into cupric sulphate solution, the above reaction occurs spontaneously. The heat of reaction ΔH for this irreversible process may be measured by calorimetric means and is found to be $-234.7\,\mathrm{kJ}$ at 298 K (the minus sign indicating an exothermic reaction). This reaction may be carried out reversibly in a Daniell cell if an opposing potential difference is applied which is infinitesimally less than the e.m.f. E of the cell, 1.0934 V. If the electrodes of the Daniell cell are connected to a motor, the work done is given by Eq. (1.14), which states

$$w = nFE$$

Since the charge on a zinc ion is equal to that of two protons, $n = 2$, the electricity transferred when 1 mole of zinc reacts is $2F$, where F is the Faraday constant.

The work done by the cell when 1 mole of zinc dissolves is

$$w = 2 \times 96\,487 \times 1.0934\,\mathrm{J}$$
$$= 208.6\,\mathrm{kJ}$$

w is the maximum work done by the system excluding expansion work, i.e. the useful work done:

$$-\Delta G = w_{useful}$$
$$\Delta G = -208.6 \text{ kJ}$$

This is the change in the Gibbs function for the reaction and is the same whether the reaction is performed reversibly or irreversibly. From Eq. (8.10),

$$T\Delta S = \Delta H - \Delta G$$
$$= -234.7 - (-208.6)$$
$$T\Delta S = -26.1 \text{ kJ}$$

Therefore 26.1 kJ is rejected by the system to the heat reservoir (constant temperature bath containing the cell) when the reaction is carried out reversibly.

Example 2   Consider the reaction

$$Pb + Hg_2Cl_2 \longrightarrow PbCl_2 + 2Hg$$

The enthalpy change for this reaction at 101 325 N m^{-2} (1 atm) and 298 K is $\Delta H^{\ominus}(298) = -105.70$ kJ. The entropy change accompanying the reaction under the same conditions is 30.67 J K^{-1}. When this reaction is carried out reversibly, $(298 \times 30.67) = 9.143$ kJ are absorbed from a heat reservoir.

The change in the Gibbs function is given by Eq. (8.10):

$$\Delta G = -105.7 - 9.143$$
$$= -114.84 \text{ kJ}$$

The maximum useful work which can be obtained if the reaction is carried out reversibly is 114.84 kJ.

8.3   Changes in the Helmholtz and Gibbs functions

Equation (8.1) states:

$$A = U - TS$$

Differentiating,

$$dA = dU - T\,dS - S\,dT \tag{8.11}$$

Employing the combination of the first and second laws (Eq. 5.16), i.e.

$$T\,dS = dU + P\,dV$$

we have

$$\boxed{dA = -P\,dV - S\,dT} \tag{8.12}$$

Equation (8.2) states:

$$G = H - TS$$

so that

$$G = U + PV - TS \tag{8.13}$$

Differentiating,

$$dG = dU + P\,dV + V\,dP - T\,dS - S\,dT \tag{8.14}$$

Employing Eq. (5.16),

$$\boxed{dG = V\,dP - S\,dT} \tag{8.15}$$

Equations (8.12) and (8.15) give the change in the Helmholtz and Gibbs functions, respectively, for a simple system as a result of any process. From these equations, it is seen that

$$\left(\frac{\partial A}{\partial V}\right)_T = -P \tag{8.16}$$

$$\left(\frac{\partial G}{\partial P}\right)_T = V \tag{8.17}$$

$$\left(\frac{\partial A}{\partial T}\right)_V = -S \tag{8.18}$$

$$\left(\frac{\partial G}{\partial T}\right)_P = -S \tag{8.19}$$

For example, let us find the change in the Gibbs function when an ideal gas passes from state 1 to state 2 during an isothermal process. From Eq. (8.15), substituting for V from the equation of state of an ideal gas, $PV = nRT$, we have, on integrating from state 1 to state 2,

$$\int_1^2 dG = nRT \int_1^2 \frac{dP}{P}$$

Hence,

$$G_2 - G_1 = nRT \ln \frac{P_2}{P_1} \qquad (8.20)$$

8.4   The Gibbs–Helmholtz equation

Let us consider a system undergoing some process such that in the final state the temperature and pressure are the same as in the initial state:

For example, this process may be a chemical reaction,

the diffusion of two ideal gases at the same temperature and pressure, or a phase change such as liquid water → ice.

Applying Eq. (8.19) to the initial and final states of the system,

$$\left(\frac{\partial G_{\text{initial}}}{\partial T} \right)_P = -S_{\text{initial}}$$

$$\left(\frac{\partial G_{\text{final}}}{\partial T} \right)_P = -S_{\text{final}}$$

Hence,

$$\left(\frac{\partial G_{\text{final}}}{\partial T} \right)_P - \left(\frac{\partial G_{\text{initial}}}{\partial T} \right)_P = -(S_{\text{final}} - S_{\text{initial}})$$

which may be written as

$$\left(\frac{\partial \Delta G}{\partial T} \right)_P = -\Delta S \qquad (8.21)$$

From Eq. (8.10),

$$\left(\frac{\partial \Delta G}{\partial T}\right)_P = \frac{\Delta G - \Delta H}{T} \tag{8.22}$$

This is the Gibbs–Helmholtz equation, which may also be written

$$\left(\frac{\partial \frac{\Delta G}{T}}{\partial T}\right)_P = -\frac{\Delta H}{T^2} \tag{8.23}$$

since the differentiation of a quotient gives

$$\left(\frac{\partial \frac{\Delta G}{T}}{\partial T}\right)_P = \frac{1}{T}\left(\frac{\partial \Delta G}{\partial T}\right)_P - \frac{\Delta G}{T^2}$$

From the Gibbs–Helmholtz equation, it is possible to find the temperature dependence of the Gibbs function. If we consider a system such that the initial and final states have the same volume and temperature, then, employing Eq. (8.18), it is possible to show in a similar manner to that above that

$$\left(\frac{\partial \Delta A}{\partial T}\right)_V = -\Delta S = \frac{\Delta A - \Delta U}{T} \tag{8.24}$$

Combining Eqs (8.21) and (8.22),

$$\left(\frac{\partial \Delta G}{\partial T}\right)_P = -\Delta S = \frac{\Delta G - \Delta H}{T} \tag{8.25}$$

Either of these equations is referred to as the Gibbs–Helmholtz equation.

Alternatively, Eq. (8.23) may be derived as follows. Remembering the differentiation of a product, we have

$$\left(\frac{\partial \frac{G}{T}}{\partial T}\right)_P = \frac{1}{T}\left(\frac{\partial G}{\partial T}\right)_P - \frac{G}{T^2} \tag{8.26}$$

Substituting from Eq. (8.19),

$$\left(\frac{\partial \frac{G}{T}}{\partial T}\right)_P = -\frac{S}{T} - \frac{G}{T^2}$$

Employing Eq. (8.2),

$$\left(\frac{\partial \frac{G}{T}}{\partial T}\right)_P = -\frac{H}{T^2} \tag{8.27}$$

This equation is also referred to as the Gibbs–Helmholtz equation. Applying the equation to the initial and final states of the system and using the fact that these states have the same temperature and pressure, Eq. (8.23) may be obtained.

8.5   The Gibbs function of an ideal gas

Equation (8.20) may be written as

$$G = G^{\ominus} + nRT \ln \frac{P}{P^{\ominus}} \tag{8.28}$$

where $G^{\ominus}$ is the Gibbs function of the system in the standard state, and $P^{\ominus}$ is the pressure in the standard state.

For an ideal gas, the standard state is defined as *the state of the pure gas at* 101 325 N m^{-2} $(= 1$ atm$)$ *pressure and the temperature under consideration.*

8.6   The Gibbs function of an ideal gas in a gaseous mixture

According to the Gibbs–Dalton law, an ideal gas in a mixture of ideal gases behaves as though it alone occupied the volume of the mixture at the same temperature. Hence, the Gibbs function, G_i, of an ideal gas i in a mixture is the same as that for the gas in the pure state at the same temperature and at its partial pressure P_i:

$$G_i = G_i^{\ominus} + n_iRT \ln \frac{P_i}{P_i^{\ominus}} \tag{8.29}$$

where $G_i^{\ominus}$ and $P_i^{\ominus}$ are the Gibbs function and pressure of gas i in the standard state.

The standard state of an ideal gas in a mixture is usually taken to be as before, that is, the state of the pure gas at $101\,325\ \text{N m}^{-2}$ and the temperature under consideration.

Equation (8.29) may be modified by applying Dalton's law (Eq. 6.8),

$$G_i = G_i^{\ominus} + n_i RT \ln \frac{x_i P}{P_i^{\ominus}} \qquad (8.30)$$

where P is the total pressure of the mixture and x_i is the mole fraction of gas i.

$$G_i = G_i^{\ominus} + n_i RT \ln \frac{P}{P_i^{\ominus}} + n_i RT \ln x_i \qquad (8.31)$$

According to the Gibbs–Dalton law, the Gibbs function of an ideal gas mixture is given by

$$G_{\text{mix}} = \Sigma_i G_i = \Sigma_i n_i g_i \qquad (8.32)$$

and the Helmholtz function by

$$A_{\text{mix}} = \Sigma_i A_i = \Sigma_i n_i a_i$$

8.7 Standard state

The standard state of a substance is taken to be the state of its stable phase at $101\,325\ \text{N m}^{-2}$ ($= 1$ atm) and the temperature under consideration. This definition is used for gases, liquids and solids (see Chapter 11 for stricter definitions). The standard Gibbs function of formation of a compound is defined as the Gibbs function change for the reaction forming 1 mole of compound from its elements, reactants and products being in the standard state. *By convention*, it is taken that *the Gibbs function for elements in their standard state is zero.* This choice is made since, with the exception of nuclear reactions, there is conservation of atoms in a chemical reaction. (Enthalpy of formation has been similarly defined.)

8.8 Entropy and energy criteria for equilibrium

Every system, if left to itself, proceeds spontaneously to a state of equilibrium. The entropy of an isolated system increases for all naturally-occurring (spontaneous) processes (see Section 5.1). However, the internal energy of an isolated system must remain constant. That is, for an isolated system,

$$\left.\begin{array}{ll} dS > 0 & \text{irreversible process} \\ dS = 0 & \text{reversible process} \\ dS < 0 & \text{impossible} \end{array}\right\} dU = 0$$

For a spontaneous process, an isolated system may only have those states with *equal internal energy* which have *higher entropy* than the initial state of the system. Therefore the system proceeds spontaneously to that state with the *maximum entropy* consistent with its initial internal energy. This is, therefore, the equilibrium state.

The equation

$$T \, dS = dU + P \, dV$$

which is true for reversible and irreversible processes, may be considered as defining a surface in a 'thermodynamic space' (Figure 8.1), with axes U, V and S. This surface is given by an equation of the form

$$S = f(U, V)$$

The plane $U = U_1$ cuts the surface. All the intersection points of the

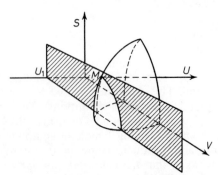

Figure 8.1. The surface formed by the possible equilibrium states of a system on a S–U–V diagram. M indicates the equilibrium state for a system of constant internal energy U_1

surface and the plane represent states which have internal energy U_1.

For all *spontaneous* processes, an *isolated system* proceeds to an *equilibrium state* — that is, to the state of *maximum entropy* which is consistent with its *initial internal energy*. The state M in Figure 8.1 therefore corresponds to the equilibrium state with internal energy U_1. So long as the system remains isolated, it will continue to have the state M, since all other states of the system with internal energy U_1 have lower entropy. Therefore *at equilibrium U and V are constant and S is a maximum.*

MATHEMATICAL APPROACH

The first law states

$$đq = dU + đw$$

Relation (5.15), which combines the first and second laws, states

$$T \, dS \geqslant dU + đw \qquad (8.33)$$

Let us consider adiabatic isochoric processes involving a simple system. For isochoric (constant volume) processes $dV = 0$ and therefore $đw = 0$. Since the process is adiabatic, $đq = 0$ and, hence, from the first law, $dU = 0$. Therefore, from relation (8.33) above,

$$\boxed{dS_{U,V} \geqslant 0} \qquad (8.34)$$

where the subscripts U and V indicate constant internal energy and volume. That is,

$$S_{\text{final}} - S_{\text{initial}} \geqslant 0$$

$$S_{\text{final}} \geqslant S_{\text{initial}}$$

Therefore the entropy tends to a maximum for an irreversible process, or at best it remains constant for a reversible process, at constant V and U. Since during an irreversible process a system proceeds to an equilibrium state, relation (8.34) is a criterion for equilibrium.

ALTERNATIVE PROOF

From relation (5.14), which states

$$dS \geqslant \frac{đq}{T}$$

we have, for any adiabatic process, since $đq = 0$,

$$\boxed{dS \geqslant 0}$$

(8.35)

$$S_{final} - S_{initial} \geqslant 0$$
$$S_{final} \geqslant S_{initial}$$

A system proceeds to an equilibrium state during an irreversible process. Hence, *the system tends to a state of maximum entropy as it approaches equilibrium during an adiabatic process.*

MECHANICAL SYSTEMS: ISOCHORIC PROCESSES

In a mechanical system there is no entropy change. For a system at constant entropy $(dS = 0)$ and constant volume $(dV = 0)$, we have, from relation (8.33),

$$\boxed{dU_{S,V} \overset{I}{\underset{R}{\leqslant}} 0}$$

Therefore the energy of a system decreases for an irreversible process. During an irreversible process, a system tends to an equilibrium state. Hence, the system tends to minimum energy as it moves toward equilibrium (see Figure 8.2).

8.9   Criteria for equilibrium: the Helmholtz and Gibbs functions

We have already seen that a system at equilibrium (1) having *constant U and V has a maximum entropy S*; (2) having *constant S and V has a minimum internal energy U*.

   In the laboratory it is difficult to construct systems of constant internal energy or entropy. In fact, we are usually only interested in systems at constant temperature and volume or at constant temperature and pressure. We shall now consider the criteria for equilibrium under these two sets of conditions.

THE HELMHOLTZ FUNCTION

At constant temperature, relation (8.4) states

$$dA \overset{I}{\underset{R}{\leqslant}} -đw$$

We shall only consider simple systems, i.e. systems in which work is done by moving boundaries alone. In such cases, at constant volume $đw = 0$

$$\boxed{\overset{\text{I}}{dA_{T,V}} \leqslant 0}_{\text{R}} \tag{8.37}$$

During a spontaneous process, a system proceeds toward equilibrium. Hence, as a system at constant temperature and volume

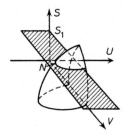

Figure 8.2. The thermodynamic surface on a S–U–V diagram cut by a constant entropy plane. N indicates the equilibrium state for a system of constant entropy S_1

proceeds to equilibrium, the Helmholtz function of the system tends to a minimum.

THE GIBBS FUNCTION

We have already seen that for an isobaric–isothermal process we have, from Eq. (8.8),

$$-dG_{P,T} = (đw_{\text{useful}})_{\text{max}}$$

Obviously, the decrease in the Gibbs function, i.e. the maximum useful work, must be greater than the useful work obtained from a system during an irreversible process,

$$-dG_{P,T} \overset{\text{I}}{\geqslant} đw_{\text{useful}} \tag{8.38}$$

If we consider *only simple* systems, then $đw_{\text{useful}} = 0$, since $đw_{\text{useful}}$ excludes the work done by these systems (i.e. by moving boundaries)

$$-dG_{P,T} \overset{\text{I}}{\geqslant} 0$$

i.e.

$$\boxed{\,^{\mathrm{I}}\!\!\!\!\underset{\mathrm{R}}{\mathrm{d}G_{P,T} \leqslant 0}\,}$$ (8.39)

$$G_{\mathrm{final}} - G_{\mathrm{initial}} \leqslant 0$$
$$G_{\mathrm{final}} \leqslant G_{\mathrm{initial}}$$

Hence, during an irreversible process at constant pressure and temperature, a simple system tends to a minimum in the Gibbs function. Since an irreversible process proceeds towards equilibrium, a system at constant pressure and temperature tends to a minimum in the Gibbs function as equilibrium is approached. For a reversible isobaric–isothermal process, there is, of course, no change in the Gibbs function for a simple system.

It is of no consequence to state that a condition for equilibrium for a system at constant temperature and pressure (i.e. $\mathrm{d}T = 0$ and $\mathrm{d}P = 0$) *is $\mathrm{d}G = 0$, since, after all, what we mean by equilibrium is that none of the thermodynamic properties of the system will change.* That is, $\mathrm{d}S = 0$, $\mathrm{d}V = 0$, $\mathrm{d}U = 0$, $\mathrm{d}H = 0$, $\mathrm{d}A = 0$, etc., will also all be true.

What is important is that (a) for a reversible, isobaric–isothermal process there is no change in the Gibbs function for a system, and (b) a system at constant temperature and pressure tends to a minimum in the Gibbs function as it approaches equilibrium.

We therefore see that when a system proceeds to equilibrium (a) *at constant V and T, the Helmholtz function A tends to a minimum;* (b) *at constant P and T, the Gibbs function G tends to a minimum.*

Relation (8.37) may be written as

$$\mathrm{d}(U - TS)_{V,T} \leqslant 0$$ (8.40)

Similarly, relation (8.39) may be written as

$$\mathrm{d}(H - TS)_{P,T} \leqslant 0$$ (8.41)

Relation (8.40) may be regarded as a combination of two opposing tendencies, (1) $\mathrm{d}U_{V,T} < 0$ and (2) $T\,\mathrm{d}S_{V,T} > 0$. It is seen that at high temperatures (2) predominates. The system tends to a state of greater disorder (i.e. solids to liquids to gases). At low temperatures (1) predominates and the system tends to a state of greater order (see Section 5.16 on disorder and entropy).

If, for a system at constant pressure and temperature, the difference in the Gibbs function, $G_2 - G_1 = \Delta G$ from state 1 to state 2, is such that ΔG is $-ve$, the system changes its state spontaneously in the direction 1→2. If ΔG is zero, the two states of the system are in equilibrium; and if ΔG is +ve, the system changes its state spontaneously in the direction 2→1 or, if the process is to proceed in the direction 1→2, work must be done on the system.

The Helmholtz function may be considered similarly for a system at constant volume and temperature.

Example   The change in the Gibbs function for the reaction

$$H_2(g) + \tfrac{1}{2}O_2(g) = H_2O(l)$$

is $-236\cdot50$ kJ mol^{-1} at 101 325 N m^{-2} and 298 K. This reaction will therefore proceed spontaneously. However, hydrogen and oxygen may be kept together at this pressure and temperature for a long time without any appreciable signs of reaction. It may be shown very simply that the reaction is spontaneous by adding a catalyst.

Thermodynamics, it must be emphasised, *is not concerned with the rates of processes*. In fact, the rate of a reaction is dependent on the concentrations of the reactants. The study of reaction rates is known as chemical kinetics.

8.10   Chemical equilibrium

All reactions attain an equilibrium between the reactants and products:

$$\text{reactants} \underset{\text{reverse}}{\overset{\text{forward}}{\rightleftarrows}} \text{products}$$

This equilibrium is attained when the forward reaction (products being formed from reactants) and the reverse reaction (reactants being formed from products) take place at the same rate. Initially, when the reactants are mixed, the forward reaction is fast. When the products are formed, the reverse reaction begins to take place and the forward reaction slows down since the concentration of the reactants has decreased. The rate of the forward reaction continues to decrease and that of the reverse reaction to increase until, at equilibrium, the two rates are equal. When this happens, there is no change in the concentration of the individual chemicals, provided that no external conditions are altered, and the system is said to be in a state of chemical equilibrium. Reactions that appear to go to completion merely have a very high product concentration relative to that of the remaining reactants. At constant pressure and temperature, a system tends to a minimum value in its Gibbs function as it approaches equilibrium. Hence, to determine the equilibrium composition for a reaction at a constant pressure and temperature, we need to evaluate the total Gibbs function for the system (i.e. reactants and products) at various stages of the reaction and determine the conditions for a minimum.

To illustrate, let us consider the water gas reaction at constant pressure, $P^{\ominus}$, and temperature 1000 K:

$$CO(g) + H_2O(g) \longrightarrow CO_2(g) + H_2(g)$$

For simplicity, we shall assume that the four substances behave as ideal gases.

If originally the reaction vessel contained 1 mole of gaseous carbon monoxide and 1 mole of water vapour and the reaction went to completion, we would find that the vessel finally contained 1 mole of carbon dioxide and 1 mole of hydrogen.

Substance	Standard molar Gibbs function $g^{\ominus}$ at 298 K (kJ mol^{-1})
CO	$-137 \cdot 180$
H_2O	$-228 \cdot 420$
CO_2	$-394 \cdot 010$
H_2	0

For an ideal gas the Gibbs function is given by Eq. (8.29), so that

$$G_{CO} = n_{CO}g_{CO}^{\ominus} + n_{CO}RT \ln \frac{P_{CO}}{P_{CO}^{\ominus}}$$

$$G_{H_2O} = n_{H_2O}g_{H_2O}^{\ominus} + n_{H_2O}RT\ln \frac{P_{H_2O}}{P_{H_2O}^{\ominus}}$$

$$G_{CO_2} = n_{CO_2}g_{CO_2}^{\ominus} + n_{CO_2} RT \ln \frac{P_{CO_2}}{P_{CO_2}^{\ominus}}$$

$$G_{H_2} = n_{H_2}g_{H_2}^{\ominus} + n_{H_2}RT \ln \frac{P_{H_2}}{P_{H_2}^{\ominus}}$$

where G_{CO} is the Gibbs function for CO in the system; $g_{CO}^{\ominus}$ is the standard molar Gibbs function for CO; P_{CO} is the partial pressure of CO, i.e. the pressure which would be observed if the CO were the only gas in the system; and n_{CO} is the number of moles of CO in the system. The other quantities are similarly defined.

Since the substances are all ideal gases, their standard pressures are all equal, $P_{H_2O}^{\ominus} = P_{CO}^{\ominus} = P_{CO_2}^{\ominus} = P_{H_2}^{\ominus} = P^{\ominus}$.

The total Gibbs function for the system is

$$G_{tot} = G_{CO} + G_{H_2O} + G_{CO_2} + G_{H_2}$$

Therefore

$$G_{tot} = G_{tot}^\ominus + RT \ln P_{CO}^{n_{CO}} \cdot P_{H_2O}^{n_{H_2O}} \cdot P_{CO_2}^{n_{CO_2}} \cdot P_{H_2}^{n_{H_2}} \cdot$$
$$\cdot (P^\ominus)^{-(n_{CO} + n_{H_2O} + n_{CO_2} + n_{H_2})} \qquad (8.42)$$

where

$$G_{tot}^\ominus = n_{CO} g_{CO}^\ominus + n_{H_2O} g_{H_2O}^\ominus + n_{CO_2} g_{CO_2}^\ominus + n_{H_2} g_{H_2}^\ominus$$

To illustrate the calculation, let us find the total Gibbs function for the system when no products are yet formed, i.e. $n_{CO_2} = n_{H_2} = 0$ and $n_{CO} = n_{H_2O} = 1$ mole.

From Eq. (6.8),

$$P_{CO} = x_{CO} P = \frac{n_{CO}}{\text{total number of moles}} \cdot P$$

Since the total pressure P is the standard pressure $P^\ominus$ in this case and there is a total of 2 moles in the system,

$$P_{CO} = \tfrac{1}{2} P^\ominus$$

Similarly, $$P_{H_2O} = \tfrac{1}{2} P^\ominus$$

Hence, $$G_{tot} = n_{CO} g_{CO}^\ominus + n_{H_2O} g_{H_2O}^\ominus + RT \ln P_{CO}^{n_{CO}} \cdot$$
$$\cdot P_{H_2O}^{n_{H_2O}} (P^\ominus)^{-(n_{CO} + n_{H_2O})}$$

$$= 1 \times (-137\ 180) + 1 \times (-228\ 420)$$

$$+ 8 \cdot 314 \times 10^3 \times 2 \cdot 303 \log_{10}(\tfrac{1}{2})^1 (\tfrac{1}{2})^1$$

$$= -377 \cdot 140 \text{ kJ}$$

As a further example of the calculation of the total Gibbs function, consider that the reaction has produced 0·6 mole of hydrogen, i.e. $n_{H_2} = n_{CO_2} = 0 \cdot 6$ mole and $n_{CO} = n_{H_2O} = 0 \cdot 4$ mole:

1 mole CO	0·6 mole CO$_2$
	0·6 mole H$_2$
1 mole H$_2$O	0·4 mole H$_2$O
	0·4 mole CO

$$P_{CO_2} = \frac{0 \cdot 6}{2} P^\ominus = 0 \cdot 3 P^\ominus$$

Similarly, $P_{H_2} = 0.3\, P^{\ominus}, P_{CO} = 0.2\, P^{\ominus}, P_{H_2O} = 0.2\, P^{\ominus}$

Substituting into Eq. (8.42),

$$G_{tot} = 0.4(-137\,180) + 0.4(-222\,420) + 0.6(-394\,010)$$

$$+ 8.314 \times 10^3 \times 2.303 \times \log_{10}(0.2)^{0.4}\,(0.2)^{0.4}\,(0.3)^{0.6}\,(0.3)^{0.6}$$

$$= -406.866\ \text{kJ}$$

The reaction will reach equilibrium when the composition of the reactants and products is such as to show a minimum in the Gibbs function. From calculations such as those made above, we can plot a graph to find the position of the minimum in the Gibbs function using the number of moles of hydrogen formed as a guide to the progress of the reaction (see Figure 8.3).

The minimum in the Gibbs function occurs when approximately 0.873 mole of hydrogen has been produced.

It must be realised that had our system originally consisted of 1 mole of CO_2 and 1 mole of H_2 under the same conditions, exactly

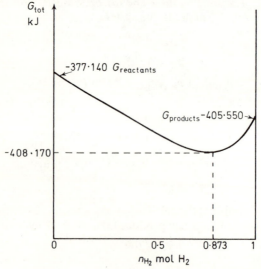

Figure 8.3. Variation of the total Gibbs function of the system with composition

the same equilibrium position would have been reached, i.e. the system at equilibrium would consist of 0.127 mole of CO, 0.127 mole of H_2O, 0.873 mole of H_2 and 0.873 mole of CO_2, since this corresponds to the position of the minimum in the total Gibbs function for the system under the stated conditions.

As we have just shown, the equilibrium position can be found by this means, but a simpler method will be shown below.

On considering the reaction at constant volume and temperature, chemical equilibrium is achieved when the total Helmholtz function for the system (reactants and products) is a minimum. Similarly, the criterion for chemical equilibrium if the reaction is carried out adiabatically is maximum entropy.

8.11 van't Hoff isotherm

In this section we determine the equilibrium position of a reaction. Consider a general chemical reaction:

$$aA + bB + \cdots \longrightarrow lL + mM + \cdots$$

where A, B, ... are reactants, L, M, ... are products and $a, b, \ldots$ and $l, m, \ldots$ are the stoichiometric coefficients required for the balance of the chemical equation. In other words, $a, b, \ldots$ are the number of moles of A, B, ..., respectively, which react to form $l, m, \ldots$ moles of L, M, ..., respectively. We shall assume that both reactants and products are ideal gases. (In Chapter 14 a more general form of the van't Hoff isotherm will be derived.)

Let the number of moles of A in the reaction vessel initially be n_A and let us similarly define $n_B, \ldots$ and $n_L, n_M, \ldots$ Consider the reaction at some point in its progress. At this point, let dn_A moles of A, dn_B moles of B, etc., react to form dn_L moles of L, dn_M moles of M, etc., so that the reaction has proceeded by an infinitesimal amount further in its course. The changes in the reactants, dn_A, dn_B, etc., must be negative quantities if the amounts of the reactants are decreasing with time.

The change in the Gibbs function due to the reaction proceeding by an infinitesimal amount is

$$dG_{tot} = g_A dn_A + g_B dn_B + \cdots + g_L\, dn_L + g_M\, dn_M + \cdots \quad (8.43)$$

where g_A, etc., are the molar Gibbs functions of the various chemical species present at the particular point of progress of the reaction under consideration.

The quantities denoted by dn are not independent of one another since the number of moles of reactants consumed and the number of moles of products formed must be proportional to the stoichiometric coefficients. That is,

$$dn_A = -a\, d\xi,\; d\, n_B = -b\, d\xi, \ldots$$
$$dn_L = l\, d\xi,\; dn_M = m\, d\xi, \ldots \quad (8.44)$$

where $d\xi$ is the constant of proportionality and ξ measures the extent of the reaction, i.e. the progress of the reaction. Initially, $\xi = 0$ and if the reaction went to completion $\xi = 1$.

Hence, employing Eq. (8.43), the change in the Gibbs function is

$$dG_{tot} = -g_A a\, d\xi - g_B b\, d\xi - \cdots + g_L l\, d\xi + g_M m\, d\xi + \cdots$$

which we shall write as

$$dG_{tot} = \Delta G\, d\xi \qquad (8.45)$$

where $$\Delta G = lg_L + mg_M + \cdots - ag_A - bg_B - \cdots \qquad (8.46)$$

ΔG is the change in the Gibbs function when stoichiometric quantities of the products are formed at the point considered in the progress of the reaction.

If the reaction is carried out at constant temperature, then, since the chemical species involved are ideal gases, we may employ Eq. (8.29). Substituting for the Gibbs functions in Eq. (8.46),

$$\Delta G = l\left(g_L^\ominus + RT \ln \frac{P_L}{P^\ominus}\right) + m\left(g_M^\ominus + RT \ln \frac{P_M}{P^\ominus}\right) + \cdots$$

$$- a\left(g_A^\ominus + RT \ln \frac{P_A}{P^\ominus}\right) - b\left(g_B^\ominus + RT \ln \frac{P_B}{P^\ominus}\right) - \cdots$$

where the P's are the partial pressures of the various chemical species at the point considered in the progress of the reaction. Hence,

$$\Delta G = lg_L^\ominus + mg_M^\ominus + \cdots - ag_A^\ominus - bg_B^\ominus - \cdots + RT \ln \left(\frac{P_L}{P^\ominus}\right)^l$$

$$+ RT \ln \left(\frac{P_M}{P^\ominus}\right)^m + \cdots - RT \ln \left(\frac{P_A}{P^\ominus}\right)^a - RT \ln \left(\frac{P_B}{P^\ominus}\right)^b - \cdots$$

Or, $$\Delta G = \Delta G^\ominus + RT \ln \frac{(P_L/P^\ominus)^l (P_M/P^\ominus)^m}{(P_A/P^\ominus)^a (P_B/P^\ominus)^b}$$

which may be written as

$$\boxed{\Delta G = \Delta G^\ominus + RT \ln \frac{(P_L)^l (P_M)^m \cdots}{(P_A)^a (P_B)^b \cdots} (P^\ominus)^{-\Delta n}} \qquad (8.47)$$

where

$$\Delta n = (l + m + \ldots) - (a + b + \ldots) \qquad (8.48)$$

and $\Delta G^\ominus$ is defined similarly to ΔG:

$$\Delta G^\ominus = lg_L^\ominus + mg_M^\ominus + \cdots - ag_A^\ominus - bg_B^\ominus - \cdots \qquad (8.49)$$

$\Delta G^\ominus$ is the change in the Gibbs function when stoichiometric quantities of the products are formed, the reactants and products being in their standard states.

If the infinitesimal reaction occurred essentially at the point of equilibrium, the partial pressures of each substance would be those at equilibrium, and the change in the total Gibbs function, $\Delta G\, d\xi$ (Eq. 8.45), would be zero. Since $d\xi$ is not zero, it being an arbitrary change in the progress of the reaction, ΔG must be zero. Hence, Eq. (8.47) becomes

$$-\Delta G^\ominus = RT \ln \left\{ \frac{(P_L)^l (P_M)^m \cdots}{(P_A)^a (P_B)^b \cdots} (P^\ominus)^{-\Delta n} \right\}_{eq} \qquad (8.50)$$

The subscript 'eq' indicates that the pressures are those at equilibrium.

At a given temperature the standard molar Gibbs function of each substance is constant. Hence, $\Delta G^\ominus$ is a constant, as also are R and T. Therefore it is seen from Eq. (8.50) that

$$\left\{ \frac{(P_L)^l (P_M)^m \cdots}{(P_A)^a (P_B)^b \cdots} (P^\ominus)^{-\Delta n} \right\}_{eq}$$

must also be a constant. This constant is known as the equilibrium constant, $K_{P/P^\ominus}$, of the reaction:

$$K_{P/P^\ominus} = \left\{ \frac{(P_L)^l (P_M)^m \cdots}{(P_A)^a (P_B)^b \cdots} (P^\ominus)^{-\Delta n} \right\}_{eq} \qquad (8.51)$$

This is a thermodynamic proof of what chemists call the *law of mass action* (partial pressures may be regarded as being proportional to the concentration of the chemical constituent).

From Eq. (8.50), we may write

$$-\Delta G^\ominus = RT \ln K_{P/P^\ominus} \qquad (8.52)$$

This equation is known as the van't Hoff isotherm.

It is seen from Eq. (8.52) that if the standard Gibbs functions for the various chemical species involved in the reaction are known, then the equilibrium constant may be calculated. Note that $K_{P/P^\ominus}$ is a dimensionless quantity (see Eq. 8.51).

Equation (8.52) may be written as

$$K_{P/P^\ominus} = e^{-\Delta G^\ominus/RT} \qquad (8.53)$$

Example 1   Calculate the change in the standard Gibbs function at 298 K for the reaction

$$N_2O_4(g) = 2NO_2(g); (K_{P/P^\ominus} = 0{\cdot}1409)$$
$$\Delta G^\ominus = -RT \ln K_{P/P^\ominus}$$
$$= -8{\cdot}314 \times 298 \times \ln 0{\cdot}1409$$
$$= -8{\cdot}314 \times 298 \times 2{\cdot}303 \times \log_{10} 0{\cdot}1409$$
$$= 4849 \text{ J}$$

Example 2   Calculate the change in the standard Gibbs function for the reaction

$$CH_4(g) + H_2O(g) = 3H_2(g) + CO(g)$$

from the following data:

	$\Delta h^\ominus(298\text{ K}) \text{ kJ mol}^{-1}$	$s^\ominus(298\text{ K}) \text{ J K}^{-1} \text{ mol}^{-1}$
$CH_4(g)$	$-74{\cdot}8$	$186{\cdot}0$
$CO(g)$	$-114{\cdot}0$	$197{\cdot}6$
$H_2(g)$	0	$132{\cdot}0$
$H_2O(g)$	$-241{\cdot}5$	$188{\cdot}6$

$\Delta G^\ominus$ may be obtained from

$$\Delta G^\ominus = \Delta H^\ominus - T\Delta S^\ominus$$
$$\Delta H^\ominus = \Delta h^\ominus(CO) + 3\Delta h^\ominus(H_2) - \{\Delta h^\ominus(CH_4) + \Delta h^\ominus(H_2O)\}$$
$$= (-114{\cdot}0) + 3 \times 0 - \{(-74{\cdot}8) + (-241{\cdot}5)\}$$
$$= 202{\cdot}3 \text{ kJ}$$
$$\Delta s^\ominus = s^\ominus(CO) + 3s^\ominus(H_2) - \{s^\ominus(CH_4) + s^\ominus(H_2O)\}$$
$$= 197{\cdot}6 + 3 \times 132{\cdot}0 - \{186{\cdot}0 + 188{\cdot}6\}$$
$$= 119{\cdot}0 \text{ J K}^{-1}$$

Hence,
$$\Delta G^\ominus = 202{\cdot}3 - 298 \times 0{\cdot}119$$
$$= 166{\cdot}8 \text{ kJ}$$

Problems

1.   One mole of ice is converted into liquid water at 273·15 K. Given that the accompanying change in volume is $1{\cdot}64 \times 10^{-6} \text{ m}^3$

mol^{-1} and that the molar latent heat of fusion is $6{\cdot}030 \text{ kJ mol}^{-1}$, calculate the change in (a) internal energy, (b) enthalpy, (c) entropy, (d) the Helmholtz function, and (e) the Gibbs function.

Answer (a) $6030{\cdot}164 \text{ J mol}^{-1}$ (b) 6030 J mol^{-1} (c) $21{\cdot}97 \text{ J K}^{-1}$ mol^{-1} (d) $-0{\cdot}164 \text{ J mol}^{-1}$ (e) 0

2. Discuss the criteria for the equilibrium of a closed system under conditions of (a) constant internal energy, (b) constant temperature and pressure, and (c) constant temperature and volume.

In what way do the properties of an isolated system change as the system approaches equilibrium?

3. Nitrogen at 298 K and 101 325 N m^{-2} is heated isobarically to 473 K. Given that $c_P(N_2) = 29{\cdot}27 \text{ J K}^{-1} \text{ mol}^{-1}$ and that its entropy at 298 K is $191{\cdot}4 \text{ J K}^{-1} \text{ mol}^{-1}$, and assuming nitrogen to behave ideally, calculate the change in (a) entropy, (b) the Helmholtz function, and (c) the Gibbs function.

Answer (a) $13{\cdot}46 \text{ J K}^{-1} \text{ mol}^{-1}$ (b) $-36{\cdot}40 \text{ kJ mol}^{-1}$ (c) $-34{\cdot}650$ kJ mol^{-1}

4. Given that the standard Gibbs function change for the formation of ammonia at 298 K is $-16{\cdot}5 \text{ kJ mol}^{-1}$, calculate the equilibrium constant of the reaction

$$N_2 + 3H_2 = 2NH_3$$

Answer 6×10^5

5. From the following data, calculate the standard entropy of solid sodium chloride:

$$Na(s) + \tfrac{1}{2}Cl_2(g) = NaCl(s)$$
$$\Delta G^{\ominus}(298{\cdot}15 \text{ K}) = -383{\cdot}3 \text{ kJ}$$
$$\Delta H^{\ominus}(298{\cdot}15 \text{ K}) = -408{\cdot}8 \text{ kJ}$$
$$s^{\ominus}(Na(s)) = 51 \text{ J K}^{-1} \text{ mol}^{-1}$$
$$s^{\ominus}(Cl_2(g)) = 104{\cdot}7 \text{ J K}^{-1} \text{ mol}^{-1}$$

Answer $72{\cdot}3 \text{ J K}^{-1}$

9
General Thermodynamic Relations

9.1 Introduction

The thermodynamic functions of state which we have employed in the discussion of closed systems in equilibrium states in the absence of electric, magnetic and gravitational fields are P, V, T, U, H, S, A and G. However, of these functions only P, V and T are easily and directly measurable properties. It is therefore desirable to derive, by the application of calculus to the laws of thermodynamics, general relationships so that from P, V and T data the remaining properties of a system may be determined.

9.2 Maxwell's equations

The combination of the first and second laws applied to a simple system, Eq. (5.16), may be written as

$$dU = T\,dS - P\,dV \qquad (9.1)$$

Since enthalpy is defined by

$$H = U + PV$$

the differential is

$$dH = dU + P\,dV + V\,dP$$

From Eq. (9.1), $\qquad dH = T\,dS + V\,dP \qquad (9.2)$

We have also previously shown that

$$dA = -P\,dV - S\,dT \qquad (9.3)$$

and $\qquad dG = V\,dP - S\,dT \qquad (9.4)$

Since U, H, A and G are functions of state, they have exact differentials which may be written in the form

$$dz = M\,dx + N\,dy \qquad (9.5)$$

and Euler's reciprocity relation must hold (see Section 1.17):

$$\left(\frac{\partial M}{\partial y}\right)_x = \left(\frac{\partial N}{\partial x}\right)_y \qquad (9.6)$$

Applying this relation to Eq. (9.1), we have

$$\boxed{\left(\frac{\partial T}{\partial V}\right)_S = -\left(\frac{\partial P}{\partial S}\right)_V} \qquad (9.7)$$

From Eq. (9.2),

$$\boxed{\left(\frac{\partial T}{\partial P}\right)_S = \left(\frac{\partial V}{\partial S}\right)_P} \qquad (9.8)$$

From Eq. (9.3),

$$\boxed{\left(\frac{\partial P}{\partial T}\right)_V = \left(\frac{\partial S}{\partial V}\right)_T} \qquad (9.9)$$

From Eq. (9.4),

$$\boxed{\left(\frac{\partial V}{\partial T}\right)_P = -\left(\frac{\partial S}{\partial P}\right)_T} \qquad (9.10)$$

These four equations are known as Maxwell's equations. Although they are simple enough to derive, it is sometimes useful to be able to write them down directly. This may be done by remembering $PV = ST$ (similar to $PV = RT$) and that *whenever P and S are in the same bracket there is a −ve sign* in the relative Maxwell's equation. If the functions of state in the brackets are cross-multiplied in any of Maxwell's equations, then $PV = ST$ is obtained. With this in mind, the four equations (9.7)–(9.10) may be written, the subscripts then being added.

Example   Show that, for an ideal gas,

$$\left(\frac{\partial U}{\partial V}\right)_T = 0, \left(\frac{\partial U}{\partial P}\right)_T = 0, \left(\frac{\partial H}{\partial V}\right)_T = 0, \left(\frac{\partial H}{\partial P}\right)_T = 0$$

Equation (9.1) states:

$$dU = T\,dS - P\,dV$$

Hence,
$$\left(\frac{\partial U}{\partial V}\right)_T = T\left(\frac{\partial S}{\partial V}\right)_T - P \tag{9.11}$$

Using Maxwell's equation (9.9), i.e.

$$\left(\frac{\partial S}{\partial V}\right)_T = \left(\frac{\partial P}{\partial T}\right)_V$$

we have

$$\left(\frac{\partial U}{\partial V}\right)_T = T\left(\frac{\partial P}{\partial T}\right)_V - P$$

For an ideal gas, $PV = nRT$. Hence, $(\partial P/\partial T)_V = P/T$.

Therefore
$$\left(\frac{\partial U}{\partial V}\right)_T = 0 \tag{9.12}$$

Similarly, from Eq. (9.2),

$$\left(\frac{\partial H}{\partial P}\right)_T = T\left(\frac{\partial S}{\partial P}\right)_T + V$$

Using Maxwell's equation (9.10), i.e.

$$\left(\frac{\partial S}{\partial P}\right)_T = -\left(\frac{\partial V}{\partial T}\right)_P$$

we have

$$\left(\frac{\partial H}{\partial P}\right)_T = -T\left(\frac{\partial V}{\partial T}\right)_P + V$$

From the equation of state for an ideal gas, $(\partial V/\partial T)_P = V/T$.

Hence,
$$\left(\frac{\partial H}{\partial P}\right)_T = 0 \tag{9.13}$$

Enthalpy is defined by $H = U + PV$. For an ideal gas, this becomes $H = U + nRT$. Hence, substituting into Eq. (9.13),

$$\left(\frac{\partial(U + nRT)}{\partial P}\right)_T = 0$$

That is,

$$\left(\frac{\partial U}{\partial P}\right)_T = 0$$

Similarly, from Eq. (9.12),

$$\left(\frac{\partial(H - nRT)}{\partial V}\right)_T = 0$$

That is,

$$\left(\frac{\partial H}{\partial V}\right)_T = 0$$

Hence, $PV = nRT$ is sufficient to define an ideal gas thermodynamically.

*9.3 The TdS equations

Entropy may be considered to be a function of temperature and volume:

$$S = f(T, V)$$

Hence,

$$dS = \left(\frac{\partial S}{\partial T}\right)_V dT + \left(\frac{\partial S}{\partial V}\right)_T dV$$

Multiplying by T,

$$T\,dS = T\left(\frac{\partial S}{\partial T}\right)_V dT + T\left(\frac{\partial S}{\partial V}\right)_T dV \tag{9.14}$$

Now, for a reversible change,

$$T\,dS = đq$$

so that

$$C_V = \left(\frac{\partial q}{\partial T}\right)_V = T\left(\frac{\partial S}{\partial T}\right)_V \tag{9.15}$$

(see Eq. 5.21).

Hence, using the Maxwell equation (9.9), Eq. (9.14) gives what is known as the *first T dS equation*:

$$T \, dS = C_V \, dT + T\left(\frac{\partial P}{\partial T}\right)_V dV \tag{9.16}$$

If the entropy of a system is now considered to be a function of temperature and pressure, then

$$dS = \left(\frac{\partial S}{\partial T}\right)_P dT + \left(\frac{\partial S}{\partial P}\right)_T dP$$

Multiplying by T,

$$T \, dS = T\left(\frac{\partial S}{\partial T}\right)_P dT + T\left(\frac{\partial S}{\partial P}\right)_T dP \tag{9.17}$$

In a similar way to that used to obtain Eq. (9.15), it can be shown that

$$C_P = T\left(\frac{\partial S}{\partial T}\right)_P \tag{9.18}$$

Hence, from Eq. (9.17) and the Maxwell equation (9.10), we obtain the *second T dS equation*:

$$T \, dS = C_P \, dT - T\left(\frac{\partial V}{\partial T}\right)_P dP \tag{9.19}$$

Employing Eq. (9.1) and the first $T \, dS$ equation (9.16), we obtain a general equation for internal energy change:

$$dU = C_V \, dT + \left[T\left(\frac{\partial P}{\partial T}\right)_V - P\right] dV \tag{9.20}$$

Employing Eq. (9.2) and the second $T \, dS$ equation (9.19), we obtain

$$dH = C_P \, dT + \left[V - T\left(\frac{\partial V}{\partial T}\right)_P\right] dP \tag{9.21}$$

Differentiating Eq. (9.20) at constant T with respect to V, we have

$$\boxed{\left(\frac{\partial U}{\partial V}\right)_T = T\left(\frac{\partial P}{\partial T}\right)_V - P} \tag{9.22}$$

This equation is known as the *energy equation* and may also be obtained by differentiating Eq. (9.1) at constant T with respect to V, which gives

$$\left(\frac{\partial U}{\partial V}\right)_T = T\left(\frac{\partial S}{\partial V}\right)_T - P$$

after which the Maxwell equation (9.9) is used.

Differentiating Eq. (9.21) at constant T with respect to P, we have

$$\boxed{\left(\frac{\partial H}{\partial P}\right)_T = V - T\left(\frac{\partial V}{\partial T}\right)_P} \tag{9.23}$$

An alternative derivation is to differentiate Eq. (9.2) with respect to P, at constant T, to give

$$\left(\frac{\partial H}{\partial P}\right)_T = T\left(\frac{\partial S}{\partial P}\right)_T + V$$

and then to employ the Maxwell equation (9.10).

Additional equations for dU and dH may be obtained in terms of other independent variables. However, these are of little practical use.

*9.4 C^P–C^V

Equating the expressions for $T\,dS$ from the first and second $T\,dS$ equations, we have

$$C_P\,dT - T\left(\frac{\partial V}{\partial T}\right)_P dP = C_V\,dT + T\left(\frac{\partial P}{\partial T}\right)_V dV$$

Rearranging, $(C_P - C_V)\,dT = T\left(\frac{\partial P}{\partial T}\right)_V dV + T\left(\frac{\partial V}{\partial T}\right)_P dP$

Dividing by dT at constant volume,

$$C_P - C_V = T\left(\frac{\partial V}{\partial T}\right)_P \left(\frac{\partial P}{\partial T}\right)_V \tag{9.24}$$

Employing the cyclic rule for partial derivatives (Eq. 1.21), we have

$$\left(\frac{\partial P}{\partial T}\right)_V = -\left(\frac{\partial V}{\partial T}\right)_P \left(\frac{\partial P}{\partial V}\right)_T$$

Hence,
$$C_P - C_V = -T\left(\frac{\partial V}{\partial T}\right)_P^2 \left(\frac{\partial P}{\partial V}\right)_T \qquad (9.25)$$

Alternatively, in terms of the thermal expansivity,

$$\alpha = \frac{1}{V}\left(\frac{\partial V}{\partial T}\right)_P$$

and the isothermal bulk modulus,

$$\beta = -V\left(\frac{\partial P}{\partial V}\right)_T$$

we have

$$C_P - C_V = VT\alpha^2\beta \qquad (9.26)$$

Equation (9.25) is important thermodynamically since it shows that:

(1) C_V may be calculated from C_P, the other partial differentials being known.
(2) C_P is never less than C_V because $(\partial V/\partial T)_P^2$ must be positive and $(\partial P/\partial V)_T$ is always negative, all substances decreasing in volume as the pressure is increased.
(3) As $T \to 0$, $C_P \to C_V$, i.e. at absolute zero, $C_P = C_V$.
(4) $C_P = C_V$ if $(\partial V/\partial T)_P = 0$. For example, for water this occurs at about 277 K when liquid water has a maximum density.

*9.5   The ratio of the heat capacities

The two $T\,\mathrm{d}S$ equations are

$$T\,\mathrm{d}S = C_P\,\mathrm{d}T - T\left(\frac{\partial V}{\partial T}\right)_P \mathrm{d}P$$

$$T\,\mathrm{d}S = C_V\,\mathrm{d}T + T\left(\frac{\partial P}{\partial T}\right)_V \mathrm{d}V$$

At constant S,

$$C_P \, dT = T\left(\frac{\partial V}{\partial T}\right)_P dP$$

$$C_V \, dT = -T\left(\frac{\partial P}{\partial T}\right)_V dV$$

Hence,

$$\gamma = \frac{C_P}{C_V} = -\left[\frac{\left(\frac{\partial V}{\partial T}\right)_P}{\left(\frac{\partial P}{\partial T}\right)_V}\right]\left(\frac{\partial P}{\partial V}\right)_S$$

Employing the cyclic rule, it is seen that the quantity in the square bracket is equal to $-(\partial V/\partial P)_T$, so that

$$\boxed{\frac{C_P}{C_V} = \left(\frac{\partial V}{\partial P}\right)_T \left(\frac{\partial P}{\partial V}\right)_S} \qquad (9.27)$$

Dividing the first and second $T \, dS$ equations by T, we obtain

$$dS = \frac{C_V}{T} \, dT + \left(\frac{\partial P}{\partial T}\right)_V dV \qquad (9.28)$$

and

$$dS = \frac{C_P}{T} \, dT - \left(\frac{\partial V}{\partial T}\right)_P dP \qquad (9.29)$$

Since entropy is a function of state, the condition for exactness, Euler's reciprocity relation (9.6), must hold. Hence, from Eq. (9.28), with $M = C_V/T$ and $N = (\partial P/\partial T)_V$,

$$\boxed{\left(\frac{\partial C_V}{\partial V}\right)_T = T\left(\frac{\partial^2 P}{\partial T^2}\right)_V} \qquad (9.30)$$

and similarly, from Eq. (9.29).

$$\boxed{\left(\frac{\partial C_P}{\partial P}\right)_T = -T\left(\frac{\partial^2 V}{\partial T^2}\right)_P} \qquad (9.31)$$

Employing the thermal expansivity α and the isothermal bulk modulus β, Eq. (9.30) may be written as

$$\left(\frac{\partial C_V}{\partial V}\right)_T = T\left(\frac{\partial(\alpha\beta)}{\partial T}\right)_V \tag{9.32}$$

and Eq. (9.31) as

$$\left(\frac{\partial C_P}{\partial P}\right)_T = -T\left(\frac{\partial(V\alpha)}{\partial T}\right)_P \tag{9.33}$$

*9.6   The Joule–Kelvin coefficient

We have already discussed the porous plug experiment in Section 2.12, where it was shown that the molar enthalpy of the gas was the same in both compartments. We shall now examine the Joule–

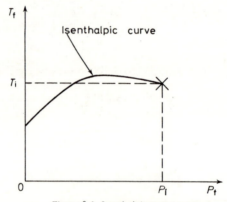

Figure 9.1. Isenthalpic curve

Kelvin effect in more detail. Consider a throttling experiment in which the pressure and temperature in the high-pressure compartment are maintained at P_i and T_i and a pressure P_f is applied in the low-pressure compartment, the temperature T_f being measured. A series of experiments is carried out in which P_i and T_i are maintained at the same values, the pressure P_f being varied and the corresponding final temperature T_f being measured. The value of the final temperature T_f is dependent on P_i, T_i and P_f. The results for a typical gas are shown in Figure 9.1, in which T_f is plotted against P_f. Points on this curve give the states of the gas which have the same molar enthalpy. The curve is known as an *isenthalpic curve*. It does not represent the actual path of the state of the gas during a throttling process since such a process is not reversible and the intermediate states cannot be shown graphically. Only the final states are shown. It is seen from Figure 9.1 that any throttling process leading to final states above the dotted horizontal line through T_i results in heating and, below this line, in cooling of the gas.

If the throttling experiment is repeated with different values for P_i and T_i, a family of isenthalpic curves may be obtained (Figure 9.2). A number of the isenthalpic curves have a state with a maximum temperature, which is known as the inversion point. The dotted curve through these points is known as the inversion curve.

The value of the slope of an isenthalpic curve at any point gives what is known as the Joule–Kelvin coefficient,

$$\mu_{JK} = \left(\frac{\partial T}{\partial P}\right)_H \tag{9.34}$$

At an inversion point $\mu_{JK} = 0$; to the left of the inversion point μ_{JK} is positive; and to the right of the inversion point μ_{JK} is negative. For states having a negative Joule–Kelvin coefficient, the result of the throttling experiment is a heating of the gas. Where μ_{JK} is positive, it is possible to obtain a cooling if the initial state is chosen correctly on a given isenthalpic. curve. A vertical line through some value of the pressure may cut the inversion curve at two different

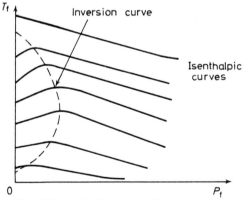

Figure 9.2. Isenthalpic curves and inversion curve

states. Therefore, for a given final pressure, there may be an upper and a lower inversion temperature.

It is seen from the inversion curve in Figure 9.2 that for the possible result of a throttling process to be a cooling, the initial temperature must be less than the maximum inversion temperature, i.e. less than the temperature at which the inversion curve cuts the temperature axis. Under normal conditions, all gases except hydrogen and helium are below their maximum inversion temperature. To liquefy hydrogen or helium by means of the Joule–Kelvin effect, the gases must be pre-cooled to below their maximum inversion temperatures.

The main importance of the Joule–Kelvin effect is that it is possible to attain low temperatures with a device having no moving parts.

Equation (9.21) states:

$$dH = C_P \, dT + \left[V - T \left(\frac{\partial V}{\partial T} \right)_P \right] dP$$

Dividing by dP while keeping the enthalpy constant, we obtain:

$$\left(\frac{\partial T}{\partial P} \right)_H = \frac{1}{C_P} \left[T \left(\frac{\partial V}{\partial T} \right)_P - V \right] = \mu_{JK} \tag{9.35}$$

Hence, μ_{JK} may be calculated for a given state from P, V, T data and a knowledge of the heat capacity for that state.

Example　Determine the Joule–Kelvin coefficient for (a) an ideal gas (b) a van der Waals' gas.

(a) The equation of state for 1 mole of an ideal gas is $Pv = RT$, so that

$$\left(\frac{\partial v}{\partial T} \right)_P = \frac{R}{P}$$

Hence, from Eq. (9.35),

$$\mu_{JK} = \frac{1}{c_P} \left[T \left(\frac{R}{P} \right) - v \right] = 0$$

That is, there is no temperature change when an ideal gas passes through a porous plug.

(b) The equation of state for a van der Waals' gas per mole is

$$\left(P + \frac{a}{v^2} \right) (v - b) = RT$$

Differentiating with respect to temperature at constant pressure, we have

$$-(v - b) \cdot \frac{2a}{v^3} \left(\frac{\partial v}{\partial T} \right)_P + \left(P + \frac{a}{v^2} \right) \left(\frac{\partial v}{\partial T} \right)_P = R$$

Employing van der Waals' equation to substitute for $(P + a/v^2)$

and rearranging, we obtain

$$\left(\frac{\partial v}{\partial T}\right)_P = \frac{Rv^3(v-b)}{RTv^3 - 2a(v-b)^2}$$

Employing this in Eq. (9.35) to find the Joule–Kelvin coefficient, we have

$$\mu_{JK} = \frac{v}{c_P}\left[\frac{RTv^2(v-b)}{RTv^3 - 2a(v-b)^2} - 1\right]$$

Or,

$$\mu_{JK} = \frac{v}{c_P}\left[\frac{2a(v-b)^2 - RTbv^2}{RTv^3 - 2a(v-b)^2}\right]$$

Problems

1. Show that the equation of state of a substance whose internal energy is dependent on temperature only is $P = Tf(V)$
2. Show that for an ideal gas C_P is independent of V and P and that the enthalpy of an ideal gas is

$$H_2 = H_1 + \int_{T_1}^{T_2} C_P \, dT$$

3. Show that for a gas obeying van der Waals' equation of state

$$du = c_V \, dT + \frac{a}{v^2} \, dv$$

and, hence, that

$$h_2 - h_1 = Pv_2 - Pv_1 - a\left(\frac{1}{v_2} - \frac{1}{v_1}\right) + \int_{T_1}^{T_2} c_V \, dT$$

4. Using any *one* of Maxwell's four equations and the cyclic rule for partial differentials, derive the remaining three.
5. Derive the following $T\, dS$ equations:

$$T \, dS = C_V \, dT + T\left(\frac{\partial P}{\partial T}\right)_V dV$$

$$T \, dS = C_P \, dT - T\left(\frac{\partial V}{\partial T}\right)_P dP$$

$$T \, dS = \frac{\kappa C_V \, dP}{\alpha} + \frac{C_P \, dV}{V\alpha}$$

where

$$\alpha = \frac{1}{V}\left(\frac{\partial V}{\partial T}\right)_P \text{ and } \kappa = -\frac{1}{V}\left(\frac{\partial V}{\partial P}\right)_T$$

6. Show that for a gas obeying van der Waals' equation the inversion temperature for a throttling process is

$$T = \frac{2a}{Rb}\left(\frac{v-b}{v}\right)^2$$

10

Open Systems

10.1 Chemical potential

The four fundamental relations which apply to closed systems, i.e. systems of constant mass, are

$$dU = T \, dS - P \, dV \tag{10.1}$$

$$dH = T \, dS + V \, dP \tag{10.2}$$

$$dA = -P \, dV - S \, dT \tag{10.3}$$

$$dG = V \, dP - S \, dT \tag{10.4}$$

We shall now extend our ideas to include open systems, i.e. systems in which mass may be transferred to or from the surroundings.

Consider a homogeneous phase in which there are a number of different substances. The internal energy U of the system may be considered to be a function of S, V and the amount of each substance present. Mathematically, this is expressed as

$$U = f(S, V, n_1, n_2, \ldots n_c)$$

where n_1, n_2, $\ldots$ n_c are the amounts (in moles) of the chemical substances $1, 2, \ldots c$. For a differential change of state,

$$dU = \left(\frac{\partial U}{\partial S}\right)_{V, n_1, n_2 \ldots n_c} . dS \quad + \quad \left(\frac{\partial U}{\partial V}\right)_{S, n_1, n_2, \ldots n_c} . dV$$

$$+ \left(\frac{\partial U}{\partial n_1}\right)_{S, V, n_2, \ldots n_c} . dn_1 \quad + \quad \left(\frac{\partial U}{\partial n_2}\right)_{S, V, n_1, n_3, \ldots n_c} . dn_2 + \cdots$$

$$\cdots + \left(\frac{\partial U}{\partial n_c}\right)_{S, V, n_1, n_2, \ldots n_c - 1} . dn_c \tag{10.5}$$

For each partial derivative, all variables other than that in the bracket are kept constant. This equation may be written as

$$dU = \left(\frac{\partial U}{\partial S}\right)_{V,n} \cdot dS + \left(\frac{\partial U}{\partial V}\right)_{S,n} \cdot dV + \Sigma_i \left(\frac{\partial U}{\partial n_i}\right)_{S,V,n'} \cdot dn_i \qquad (10.6)$$

where the subscript n on the first two partial derivatives indicates that the amount of every substance is kept constant (i.e. as for a closed system). The subscript n' indicates that the amount of every substance except the one in the bracket being considered is kept constant (see Eq. 10.5). It is seen, on comparing Eq. (10.6) with Eq. (10.1) for closed systems, that

$$\left(\frac{\partial U}{\partial S}\right)_{V,n} = T \text{ and } \left(\frac{\partial U}{\partial V}\right)_{S,n} = -P$$

Equation (10.6) may be written as

$$\boxed{dU = T\,dS - P\,dV + \Sigma_i \mu_i\,dn_i} \qquad (10.7)$$

where

$$\boxed{\mu_i = \left(\frac{\partial U}{\partial n_i}\right)_{S,V,n'}}$$

μ_i is known as the *chemical potential* of substance i.

$$\boxed{dU = T\,dS - P\,dV + \mu_1\,dn_1 + \mu_2\,dn_2 + \cdots}$$

Chemical potential μ_i is an intensive property. It may be interpreted physically as the rate of change of the internal energy per mole of substance i when an infinitesimal quantity of i is added to the system, S, V and the amounts of all other substances in the system being kept constant.

Enthalpy is defined by $H = U + PV$. Hence,

$$dH = dU + P\,dV + V\,dP$$

Substituting for dU from Eq. (10.7), we have

$$\boxed{dH = T\,dS + V\,dP + \Sigma_i \mu_i\,dn_i} \qquad (10.8)$$

The Helmholtz function is defined by $A = U - TS$. Hence,

$$dA = dU - T\,dS - S\,dT$$

Employing Eq. (10.7), we have

$$dA = -S\,dT - P\,dV + \Sigma_i\,\mu_i\,dn_i \qquad (10.9)$$

The Gibbs function is defined by $G = H - TS = U + PV - TS$.

Hence, $\qquad dG = dU + P\,dV + V\,dP - SdT - T\,dS$

Employing Eq. (10.7), we have

$$dG = V\,dP - S\,dT + \Sigma_i\,\mu_i\,dn_i \qquad (10.10)$$

It can be seen from Eqs (10.7)–(10.10) that

$$\mu_i = \left(\frac{\partial U}{\partial n_i}\right)_{S,V,n'} = \left(\frac{\partial H}{\partial n_i}\right)_{S,P,n'} = \left(\frac{\partial A}{\partial n_i}\right)_{T,V,n'} = \left(\frac{\partial G}{\partial n_i}\right)_{T,P,n'} \qquad (10.11)$$

Chemical potential may be given physical interpretations, as before, in terms of H, A and G. For example, the chemical potential of the species i is the rate of change of the Gibbs function per mole of i when an infinitesimal quantity of i is added to the system at constant pressure, temperature and composition of the system. Or a simpler way of regarding the chemical potential of species i is as the molar Gibbs function of i in the environment of the system being considered.

Equation (10.10) could have been obtained by considering the Gibbs function to be given by an equation

$$G = f(P, T, n_1, n_2, \ldots n_c)$$

The total differential of G as given by this equation is found to be equivalent to Eq. (10.10) on comparison with Eq. (10.4).

10.2 Partial molar properties

For a single-phase system, any extensive function of state Z (e.g. V, U, H, S, A, G, etc.) may be considered as a function of temperature, pressure and the amount of each chemical species in the system. That is,

$$Z = f(T, P, n_1, n_2, \ldots n_c) \qquad (10.12)$$

The total differential is given by

$$dZ = \left(\frac{\partial Z}{\partial T}\right)_{P,n} dT + \left(\frac{\partial Z}{\partial P}\right)_{T,n} dP + \sum_i \left(\frac{\partial Z}{\partial n_i}\right)_{T,P,n'} dn_i \qquad (10.13)$$

Let

$$\boxed{\bar{Z}_i = \left(\frac{\partial Z}{\partial n_i}\right)_{T,P,n'}} \qquad (10.14)$$

$\bar{Z}_i$ is known as the partial molar property of species i in the mixture, and is an intensive property. For a one-component system (i.e. for a pure substance), Eq. (10.14) reduces to $\bar{Z}_i = Z/n = z$, where z is the molar property of the system and n is the amount in moles of the component. The partial molar property $\bar{Z}_i$ of species i is the rate of change of the extensive property Z per mole when an infinitesimal amount of species i is added to the system, the pressure, temperature and amounts of all other species being constant. $\bar{Z}_i$ is the molar property Z of species i in the environment of the system being considered.

It is seen, on comparing Eqs (10.11) and (10.14), that the chemical potential μ_i is the partial molar Gibbs function of species i.

Equation (10.13) may be written as

$$dZ = \left(\frac{\partial Z}{\partial T}\right)_{P,n} dT + \left(\frac{\partial Z}{\partial P}\right)_{T,n} dP + \sum_i \bar{Z}_i \, dn_i$$

or, at constant T and P,

$$\boxed{dZ = \sum_i \bar{Z}_i \, dn_i} \qquad (10.15)$$

Consider a system consisting of a single phase. Any extensive property Z of the system — for example, volume V, internal energy U, enthalpy H, entropy S, Gibbs' function G, etc. — may be expressed by Eq. (10.12). Remember extensive properties depend on the amount of material present. At equilibrium, the components of the system are evenly distributed throughout.

Imagine that an infinitesimal portion of the system is taken away, so that the change in the amount of component i is given by

$$dn_i = -n_i \, dk \qquad (10.16)$$

where dk is a proportionality factor which is dependent on the amount of the system taken away and the minus indicates a decrease. The infinitesimal changes in some of the other extensive properties are $dV = -V \, dk, dU = -U \, dk, dS = -S \, dk, dG = -G \, dk$ (taking half the system away would halve the volume) and, in general, any extensive property Z of the system is changed by

$$dZ = -Z\,dk \qquad (10.17)$$

The intensive state (P, T) of the system is the same as that of the original system. Using Eqs (10.16) and (10.17) to substitute into Eq. (10.15), we have

$$Z\,dk = \Sigma_i \bar{Z}_i n_i\,dk$$

Hence,

$$\boxed{Z = \Sigma_i n_i \bar{Z}_i} \qquad (10.18)$$

Since Z may be any extensive function of state, we have, for example,

$$U = \Sigma_i n_i \bar{U}_i,\ H = \Sigma_i n_i \bar{H}_i,\ S = \Sigma_i n_i \bar{S}_i,\ V = \Sigma_i n_i \bar{V}_i,\ G = \Sigma_i n_i \mu_i$$

The concept of partial molar quantities has been devised so that the properties of a real mixture may be evaluated, since the molar properties of the pure components of a mixture cannot be used except when the mixture is ideal. For example, it is well known that the volume obtained on mixing two miscible liquids is frequently not the sum of the volumes of the two separate liquids. The volume of an ethanol–water mixture is less than the sum of the volumes of the unmixed liquids. At a given temperature and pressure, this deviation varies with the composition of the mixture.

Example It is known that at 298 K the density of ethanol is $785 \cdot 1\ \mathrm{kg\ m^{-3}}$ and that of water is $997 \cdot 1\ \mathrm{kg\ m^{-3}}$. Also, the partial molar volumes of ethanol and water are, respectively, $57 \cdot 30 \times 10^{-6}$ and $17 \cdot 15 \times 10^{-6}\ \mathrm{m^3\ mol^{-1}}$ in an ethanol–water mixture which is 60% ethanol by weight. Determine the volume of 1 kg of mixture assuming that (a) the mixture is ideal, i.e. employing the densities, (b) the mixture is not ideal, i.e. employing the partial molar volumes. The molecular weights of ethanol and water are 46·07 and 18·02, respectively.

(a) The volume of the mixture is the sum of the volumes of the pure components,

$$\text{volume} = \frac{0 \cdot 6}{785 \cdot 1} + \frac{0 \cdot 4}{997 \cdot 1} = 1 \cdot 1656 \times 10^{-3}\ \mathrm{m^3}$$

(b) In 1 kg of mixture,

the number of moles of ethanol, $n_{\mathrm{EtOH}} = \dfrac{600}{46 \cdot 07} = 13 \cdot 02\ \mathrm{mol}$

the number of moles of water, $n_{\mathrm{H_2O}} = \dfrac{400}{18 \cdot 02} = 22 \cdot 20\ \mathrm{mol}$

The volume of the mixture is given by

$$\text{volume} = \Sigma_i n_i \bar{V}_i = n_{EtOH}\bar{V}_{EtOH} + n_{H_2O}\bar{V}_{H_2O}$$
$$= 13\cdot02 \times 57\cdot30 \times 10^{-6} + 22\cdot20 \times 17\cdot15 \times 10^{-6}$$
$$= 1\cdot1269 \times 10^{-3} \text{ m}^3$$

10.3 The Gibbs–Duhem equation

Let the extensive property in Eq. (10.18) be the Gibbs function, so that

$$G = \Sigma_i n_i \mu_i$$

Differentiating,

$$dG = \Sigma_i n_i \, d\mu_i + \Sigma_i \mu_i \, dn_i$$

Comparing this equation with Eq. (10.10), it is seen that

$$\boxed{V \, dP - S \, dT - \Sigma_i n_i \, d\mu_i = 0} \qquad (10.19)$$

This is known as the *Gibbs–Duhem Equation*. It may be applied to any homogeneous phase.

It is seen that if there are c independently variable components in a phase, then there are $c+1$ independent intensive properties, i.e. $c+1$ properties are required to define the state of a system.

We see from the Gibbs–Duhem equation that

$$\boxed{\text{at constant } T \text{ and } P, \quad \Sigma_i n_i \, d\mu_i = 0} \qquad (10.20)$$

A more general equation may be obtained as follows. Differentiating Eq. (10.18), we have the change in the property Z given by

$$dZ = \Sigma_i \bar{Z}_i \, dn_i + \Sigma_i n_i \, d\bar{Z}_i \qquad (10.21)$$

Subtracting Eq. (10.15) from Eq. (10.21), we have the Gibbs–Duhem equation at constant T and P:

$$\boxed{\Sigma_i n_i \, d\bar{Z}_i = 0} \qquad (10.22)$$

Or, dividing by the total number of moles in the system given by

$$N = \Sigma_i n_i$$

since the mole fraction of component i is $x_i = n_i/N$, Eq. (10.22) becomes

$$\Sigma_i x_i \, d\bar{Z}_i = 0 \qquad (10.23)$$

For a two-component, single-phase system at constant T and P, we therefore have

$$x_A \, d\bar{Z}_A + x_B \, d\bar{Z}_B = 0 \qquad (10.24)$$

Dividing by dx_A,

$$x_A \left(\frac{\partial \bar{Z}_A}{\partial x_A}\right)_{P,T} + x_B \left(\frac{\partial \bar{Z}_B}{\partial x_A}\right)_{P,T} = 0 \qquad (10.25)$$

If Z were the Gibbs function, then for a two-component system we would have

$$x_A \left(\frac{\partial \mu_A}{\partial x_A}\right)_{P,T} + x_B \left(\frac{\partial \mu_B}{\partial x_A}\right)_{P,T} = 0 \qquad (10.26)$$

10.4 Chemical potential and mass transfer between phases

Consider a system consisting of two phases A and B which are in thermal equilibrium. The system as a whole is closed to mass transfer

Figure 10.1.

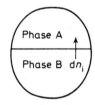

Figure 10.2.

but mass transfer may occur between the phases. For example, consider the system in Figure 10.1 where iodine may be transferred between the water and carbon tetrachloride layers.

Let an infinitesimal amount dn_i moles of component i pass from phase B to phase A, Fig. 10.2.

The pressure of the two phases need not be the same if they are separated by a rigid semipermeable membrane, the membrane being permeable to some components but not to others. For example, in an osmotic pressure experiment (Figure 10.3), the membrane is permeable only to the solvent molecules. The solvent may be transferred between the solution and the pure solvent.

The change in the internal energy of each phase due to the mass transfer of dn_i moles of component i is given by Eq. (10.7):

$$dU^A = T\,dS^A - P^A\,dV^A + \mu_i^A\,dn_i$$

$$dU^B = T\,dS^B - P^B\,dV^B - \mu_i^B\,dn_i$$

where the superscripts refer to a particular phase and the last term

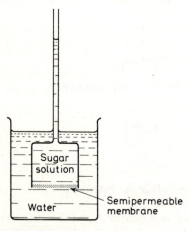

Figure 10.3. Osmotic pressure

in the second equation is negative because the mass is transferred from phase B to phase A.

On adding these two equations, we obtain the total energy change of the system. If we denote the work done by $đw$, so that

$$đw = P^A\,dV^A + P^B\,dV^B$$

we have

$$dU = T\,dS - đw + (\mu_i^A - \mu_i^B)\,dn_i \qquad (10.27)$$

For a closed system, the first law may be written as $dU = đq - đw$.

Employing this in Eq. (10.27) and rearranging,

$$(\mu_i^B - \mu_i^A)\,dn_i = T\,dS - đq$$

Now consider that the changes need not be reversible. The second law (Eq. 5.14) requires that

$$T\,dS - đq \underset{R}{\overset{I}{\geqslant}} 0$$

Hence,

$$\boxed{\mu_i^B - \mu_i^A \geqslant 0} \qquad (10.28)$$

During a reversible process, the system passes through a series of equilibrium states. Therefore, the equality applies when the two phases are in equilibrium,

$$\mu_i^A \stackrel{R}{=} \mu_i^B \qquad (10.29)$$

Hence, *if the two phases are in equilibrium, the chemical potential of component i must be the same in each phase.* This is the criterion for phase equilibrium.

For example, in the $H_2O-I_2-CCl_4$ system in Figure 10.1, the two phases would be in equilibrium if the chemical potential of iodine in the water phase were the same as that in the carbon tetrachloride phase; no further changes would occur in the iodine concentration. The inequality applies to irreversible, i.e. spontaneous, processes.

$$\mu_i^B \stackrel{I}{>} \mu_i^A$$

That is, mass transfer occurs spontaneously from a region of higher to a region of lower chemical potential. This is why the term 'chemical potential' is employed, since it may be regarded as the driving force behind mass transfer. The diffusion of a component through a system is due to chemical potential differences that occur within the system for that component.

For example, consider the system given in Figure 10.3. The water is spontaneously transferred through the semipermeable membrane owing to the existence of a difference in the chemical potential of water on either side of the membrane. This transfer will continue until the chemical potentials of the water in the water phase and in the solution are equal:

$$\mu_{H_2O}^{\text{solution}} = \mu_{H_2O}^{H_2O}$$

10.5 Relationships between partial molar properties

It is seen from Eq. (10.10) that

$$\left(\frac{\partial G}{\partial P}\right)_{T,n} = V \qquad (10.30)$$

The partial molar volume of a component i in a system is defined by Eq. (10.14)

$$\bar{V}_i = \left(\frac{\partial V}{\partial n_i}\right)_{T,P,n'} \qquad (10.31)$$

From these two equations and the definition of chemical potential, we have

$$\left(\frac{\partial \mu_i}{\partial P}\right)_{T,n} = \frac{\partial}{\partial P}\left(\frac{\partial G}{\partial n_i}\right)_{T,P,n'} = \frac{\partial}{\partial n_i}\left(\frac{\partial G}{\partial P}\right)_{T,n} = \left(\frac{\partial V}{\partial n_i}\right)_{T,P,n'} = \bar{V}_i$$

(employing the mathematical fact that the order of partial differentiation is immaterial).

That is,

$$\left(\frac{\partial \mu_i}{\partial P}\right)_{T,n} = \bar{V}_i \qquad (10.32)$$

It is also seen from Eq. (10.10) that

$$\left(\frac{\partial G}{\partial T}\right)_{P,n} = -S \qquad (10.33)$$

Hence, we obtain similarly

$$\left(\frac{\partial \mu_i}{\partial T}\right)_{P,n} = \frac{\partial^2 G}{\partial T \partial n_i} = \frac{\partial^2 G}{\partial n_i \partial T} = -\left(\frac{\partial S}{\partial n_i}\right)_{T,P,n'} = -\bar{S}_i$$

That is,

$$\left(\frac{\partial \mu_i}{\partial T}\right)_{P,n} = -\bar{S}_i \qquad (10.34)$$

As a general rule, all the thermodynamic relations which have been previously derived for closed (molar) systems may be applied to partial molar properties. All extensive properties in those equations must be replaced by the corresponding partial molar property. Compare Eq. (10.32) with Eq. (10.30), which is, of course, Eq. (8.17). Also, compare Eq. (10.34) with Eq. (10.33), which has also been derived previously, Eq. (8.19). As further examples, we may write:

from Eq. (8.2), $\mu_i = \bar{H}_i - T\bar{S}_i$ (10.35)

from Eq. (8.27), $\left(\dfrac{\partial(\mu_i/T)}{\partial T}\right)_{P,n} = -\dfrac{\bar{H}_i}{T^2}$ (10.36)

from Eq. (2.24), $\left(\dfrac{\partial \bar{H}_i}{\partial T}\right)_P = \bar{C}_{P_i}$

From Eq. (8.29), for 1 mole of an ideal gas we have

$$g_i = g_i^{\ominus} + RT \ln \frac{P_i}{P_i^{\ominus}}$$

Hence, substituting partial molar properties in this equation for an ideal gas i,

$$\mu_i = \mu_i^\ominus + RT \ln \frac{P_i}{P_i^\ominus} \qquad (10.37)$$

10.6 Chemical potential in terms of the intensive properties P, T, x

Consider a homogeneous two-component system. For a component A, the chemical potential, which must be expressed in terms of intensive variables since it is itself intensive, may be expressed as a function of pressure, temperature and the mole fraction of A:

$$\mu_A = f(P, T, x_A)$$

Thus

$$d\mu_A = \left(\frac{\partial \mu_A}{\partial P}\right)_{T,n} dP + \left(\frac{\partial \mu_A}{\partial T}\right)_{P,n} dT + \left(\frac{\partial \mu_A}{\partial x_A}\right)_{P,T} dx_A$$

Employing Eqs (10.32) and (10.34),

$$d\mu_A = \bar{V}_A\, dP - \bar{S}_A\, dT + \left(\frac{\partial \mu_A}{\partial x_A}\right)_{P,T} dx_A \qquad (10.38)$$

A similar equation holds for component B.

For a multicomponent system, the differential of the chemical potential of a component may be derived similarly by expressing the chemical potential in terms of P, T and $c-1$ mole fractions, giving

$$d\mu_i = \bar{V}_i\, dP - \bar{S}_i\, dT + \sum_i^{c-1} \left(\frac{\partial \mu_i}{\partial x_i}\right)_{P,T} dx_i$$

where the summation is over $c-1$ components.

10.7 Determination of partial molar quantities

The methods for the determination of the various partial molar quantities, $\bar{V}_i$, $\bar{H}_i$, $\bar{S}_i$, μ_i, etc., are all exactly similar and therefore, in the following, we shall refer to a general property Z which may be any of the extensive properties V, H, S, G, etc.

(1) If Z is known as a function of n_1, n_2, ..., then by partial differentiation the partial molar property of component i, $\bar{Z}_i$, may be calculated (see Eq. 10.14).

(2) From experimental data, Z is measured by varying the amount of component 1, keeping the amounts of all the other components in the solution constant. Since

$$\left(\frac{\partial \dfrac{Z}{n_1}}{\partial \dfrac{}{n_2+n_3+...}} \right)_{n_2, n_3, ...} = (n_2+n_3+...)\left(\frac{\partial Z}{\partial n_1} \right)_{n_2, n_3, ...}$$

$$= (n_2+n_3+...)\bar{Z}_1$$

if Z is plotted against the mole ratio $n_1/(n_2+n_3+...)$, then, at a given composition, $\bar{Z}_1$ is equal to the gradient at that point divided by $(n_2+n_3+...)$.

For binary mixtures, it is convenient to use molality as a measure of concentration. The molality of A, m_A, is the number of moles of A in 1 kg of B. In solutions of different molality of A, the quantity of B is constant, i.e. 1 kg. If the extensive property Z of the solution is plotted against the molality of A, then $\bar{Z}_A$ is equal to the gradient at the molality of interest divided by $n_B = 1000/$molecular weight of B.

(3) *Intercept method*   This method is applicable to two-component solutions.

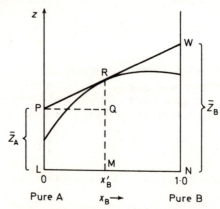

Figure 10.4. Variation of a molar property with mole fraction

The molar property, $z = Z/(n_A+n_B)$, is plotted against the mole fraction (Figure 10.4). The tangent to the curve at a particular composition has intercepts at $x_A = 1$ and $x_B = 1$ which are equal to $\bar{Z}_A$ and $\bar{Z}_B$, respectively.

For an isothermal, isobaric change in a solution of A and B, Eq. (10.15) may be written as

$$dZ = \bar{Z}_A \, dn_A + \bar{Z}_B \, dn_B$$

Dividing both sides of the equation by the total number of moles, $n_A + n_B$,

$$dz = \bar{Z}_A \, dx_A + \bar{Z}_B \, dx_B$$

For a binary solution, $x_A + x_B = 1$, so that we have

$$dx_A + dx_B = 0$$

Therefore

$$dz = \bar{Z}_A(-dx_B) + \bar{Z}_B \, dx_B$$

That is,

$$\frac{dz}{dx_B} = \bar{Z}_B - \bar{Z}_A$$

Multiplying throughout by x_B,

$$x_B \frac{dz}{dx_B} = x_B \bar{Z}_B - x_B \bar{Z}_A$$

Or,

$$x_B \frac{dz}{dx_B} = x_B \bar{Z}_B + x_A \bar{Z}_A - \bar{Z}_A \tag{10.39}$$

However, on dividing Eq. (10.18) by $n_A + n_B$ throughout, we have that

$$z = x_A \bar{Z}_A + x_B \bar{Z}_B$$

so that substituting into Eq. (10.39) and rearranging gives

$$\bar{Z}_A = z - x_B \frac{dz}{dx_B}$$

From Figure 10.4,

$$\bar{Z}_A = RM - LM.\frac{RQ}{PQ} = RM - RQ$$

Therefore

$$\bar{Z}_A = PL$$

Similarly, it can be shown that $\bar{Z}_B = WN$.

(4) *Use of the Gibbs–Duhem equation.* This method is applicable to binary mixtures when the variation of a partial molar property with concentration is known for one component. The method permits the calculation of $\bar{Z}$ for the other component and can thus be used to check the results of other methods.

For an isothermal, isobaric change in a two-component solution, Eq. (10.24) states

$$x_A \, d\bar{Z}_A + x_B \, d\bar{Z}_B = 0$$

Rearranging, $$d\bar{Z}_A = -\frac{x_B}{x_A} \, d\bar{Z}_B$$

$$d\bar{Z}_A = -\frac{x_B}{1 - x_B} \, d\bar{Z}_B$$

Integrating from $x_B = 0$ (i.e. pure A) to the state required, noting that the partial molar property of A for the pure state is the molar property z_A,

$$\int_{z_A}^{z_A} d\bar{Z}_A = -\int_{x_B = 0}^{x_B} \frac{x_B}{1 - x_B} \, d\bar{Z}_B$$

$$\bar{Z}_A = z_A - \int_{x_B = 0}^{x_B} \frac{x_B}{1 - x_B} \, d\bar{Z}_B$$

Therefore, if $x_B/(1 - x_B)$ is plotted against $\bar{Z}_B$, the integral may be obtained by measuring the area under the curve from $x_B = 0$ to x_B. Hence, $\bar{Z}_A$ may be found.

Problems

1.   From the table given below, calculate the volume of the following solutions:
(a) 2 moles water with 8 moles ethanol;   (b) 9 moles water with 1 mole ethanol;   (c) 6 moles water with 4 moles ethanol. What is the total volume change when solutions (a) and (c) are mixed?

Mole per cent of ethanol	$\bar{V}_{water}$ $\times 10^{-6} \, m^3 \, mol^{-1}$	$\bar{V}_{ethanol}$ at 293 K
0	18·05	54·20
10	18·11	53·10
20	17·67	55·40
40	17·01	57·10
60	16·21	57·97
80	15·37	58·30

Answer   (a) $0.497 \, 16 \times 10^{-3} \, m^3$, (b) $0.216 \, 09 \times 10^{-3} \, m^3$, (c) $0.330 \, 46 \times 10^{-3} \, m^3$. Volume change $= -2.3 \times 10^{-6} \, m^3$.

2.   Repeat the calculations of problem 1 above assuming the solutions to be ideal.

Answer   (a) $0.470 \, 70 \times 10^{-3} \, m^3$, (b) $0.215 \, 55 \times 10^{-3} \, m^3$, (c) $0.325 \, 10 \times 10^{-3} \, m^3$. No volume change.

11

One-Component Systems

11.1 The chemical potential of ideal gases

We have already stated that an ideal gas has the equation of state $PV = nRT$, or, for 1 mole, $Pv = RT$. For a pure substance, i.e. a one-component system, a partial molar property is the same as the molar property (see Section 10.2). Hence, for a system consisting of a pure ideal gas, the partial molar volume is the same as the molar volume, $\bar{V} = v$, and $\mu = g$. However, for simplicity we shall retain the symbol μ in the latter case, since when dealing with solutions we shall need to use partial molar quantities and thus be concerned with chemical potential.

Applying Eq. (10.32) or Eq. (8.17) to an ideal-gas system

$$\left(\frac{\partial \mu}{\partial P}\right)_T = v \tag{11.1}$$

Hence, at constant T,

$$d\mu = \frac{RT}{P} dP = RT \, d \ln P \tag{11.2}$$

Integrating from state 1 to state 2,

$$\int_{\mu_1}^{\mu_2} d\mu = \int_{P_1}^{P_2} RT \, d \ln P$$

$$\mu_2 - \mu_1 = RT \ln \frac{P_2}{P_1}$$

Let state 1 be a reference state, i.e. the standard state. Then we have

$$\mu - \mu^{\ominus} = RT \ln \frac{P}{P^{\ominus}} \tag{11.3}$$

where $P^{\ominus}$ and $\mu^{\ominus}$ are the pressure and chemical potential, respectively, in the standard state. This equation may also be obtained from Eq. (8.28) on dividing through by n.

Since chemical potential is an energy function and is therefore not defined explicitly (i.e. it is defined in terms of an arbitrary constant), its value is dependent on the definition of the standard state.

Normally, we define *the standard state of an ideal gas as the state of the pure gas at* $101\,325\,\text{N m}^{-2}$ ($= 1$ atm) *pressure and at the temperature under consideration*. Hence, the value of $\mu^{\ominus}$ depends only on the temperature, since the pressure is constant ($101\,325\,\text{N m}^{-2}$). Obviously, if the gas is in its standard state, then $P = P^{\ominus}$ and, from Eq. (11.3), the chemical potential is then the standard chemical potential.

11.2   The effect of total pressure on vapour pressure

Consider a system consisting of a pure liquid in equilibrium with its vapour at a temperature T. Let the vapour pressure be p. An inert gas is now introduced isothermally into the system so as to increase the total pressure by dP. We shall assume that the solubility of the gas in the liquid is negligible and that the gas and vapour behave ideally.

The vapour pressure of the liquid will be different in the presence of the gas. Since before and after the introduction of the gas the liquid and vapour were in equilibrium, then, according to Eq. (10.29),

$$d\mu_{\text{liq}} = d\mu_{\text{vap}} \qquad (11.4)$$

Now $\qquad\qquad d\mu_{\text{liq}} = v_{\text{liq}}\, dP - s_{\text{liq}}\, dT$

However, since the process is isothermal,

$$d\mu_{\text{liq}} = v_{\text{liq}}\, dP$$

Since the vapour behaves ideally, we have, from Eq. (11.2), that

$$d\mu_{\text{vap}} = RT\, d \ln p$$

Therefore Eq. (11.4) gives

$$v_{\text{liq}}\, dP = RT\, d \ln p$$

Integrating at constant temperature from the state where no gas was present (the pressure being merely that due to the vapour p) to the state where, owing to the introduction of the gas, the total pressure was P and the vapour pressure is altered to p',

$$\int_{p}^{P} v_{\text{liq}} \, dP = RT \int_{p}^{p'} d \ln p$$

Since liquids are almost incompressible under normal conditions, the molar volume v_{liq} is very nearly constant. Hence,

$$\ln \frac{p'}{p} = \frac{v_{\text{liq}}(P-p)}{RT} \qquad (11.5)$$

Since the right-hand side must be positive, the vapour pressure of a liquid increases as the total pressure is increased.

Example At 293 K the vapour pressure of water is $565 \cdot 8 \text{ N m}^{-2}$. Calculate the vapour pressure of water at a total pressure of 202 650 N m^{-2}, given that the molar volume of water is $1 \cdot 8 \times 10^{-5} \text{ m}^3$.

$$\ln \frac{p'}{565 \cdot 8} = \frac{1 \cdot 8 \times 10^{-5} \times 202\,650}{8 \cdot 314 \times 293}$$

$$p' = 565 \cdot 8 \text{ N m}^{-2}$$

It may be deduced from the above example that the vapour pressure of a liquid hardly increases at all with total pressure, i.e. p'/p is very nearly unity, except at very high pressure.

11.3 Real gases

The properties of real gases may be evaluated from (a) P, V, T data, (b) an equation of state, (c) the law of corresponding states.

Many equations of state have been proposed for real gases. A few of the better known, given on the mole basis, are:

van der Waals'

$$\left(P + \frac{a}{v^2}\right)(v - b) = RT \qquad (11.6)$$

where a/v^2 and b are correction terms for the intermolecular forces and for the finite volume of the molecules, respectively.

Dieterici

$$P(v - b)e^{a/vRT} = RT$$

Beattie–Bridgman

$$Pv^2 = RT\left\{v + B_0\left(1 - \frac{b}{v}\right)\right\}\left(1 - \frac{c}{vT^3}\right) - A_0\left(1 - \frac{a}{v}\right)$$

where A_0, B_0, a, b, c are constants.

The virial equations of state, in which Pv is expressed as a function of P or v, are

$$Pv = RT + BP + CP^2 + DP^3 + \cdots$$

and

$$Pv = RT\left(1 + \frac{B'}{v} + \frac{C'}{v^2} + \frac{D'}{v^3} + \cdots\right)$$

where B, B', C, C', etc., are known as the virial coefficients and are functions of temperature only. The virial equations may be made to agree with experimental data as closely as desired by increasing the number of coefficients.

It is sometimes convenient to write

$$Pv = zRT \tag{11.7}$$

where z is known as the compressibility factor and depends on both temperature and pressure. Obviously, for an ideal gas $z = 1$ for all temperatures and pressures. The compressibility factor is a measure of the deviation of a gas from ideal behaviour. A real gas approaches ideal behaviour ($z \rightarrow 1$) as its pressure is decreased ($P \rightarrow 0$) at a given temperature.

THE PRINCIPLE OF CORRESPONDING STATES

All fluids in the same reduced state have approximately equal values for their thermodynamic properties. The reduced state is specified by two reduced properties such as the reduced pressure, P_R, reduced temperature, T_R, and reduced molar volume, v_R. A reduced property is defined as the actual property divided by the value of that property in the critical state, e.g. $P_R = P/P_c$, $T_R = T/T_c$, $v_R = v/v_c$, where P_c, T_c, and v_c are the critical pressure, temperature and molar volume, respectively.

On the basis of this principle, *all gases in the same reduced state deviate from ideal gas behaviour to approximately the same extent.* The deviation from ideal behaviour may be shown by plotting the compressibility factor z against the reduced pressure for various reduced temperatures T_R as in Figure 11.1, where data for a large number of gases have been used to form a general compressibility chart. The average error in the chart compared with experimental data for most gases is less than 3 %.

From Figure 11.1, it can be seen that as the pressure is decreased

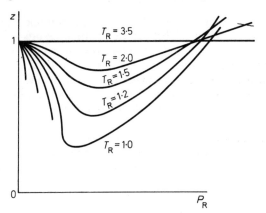

Figure 11.1. Graph of compressibility against reduced pressure for various reduced temperatures, the results from a large number of gases being used to obtain average values

($P \to 0$) all real gases tend to ideal behaviour ($z \to 1$). It is possible to calculate all the properties of real gases by making use of Figure 11.1.

11.4 The chemical potential of real gases

It is obvious that Eq. (11.2), which states that for an isothermal change

$$d\mu = RT \, d \ln P$$

does not apply to real gases. In order to evaluate the change in the chemical potential for an isothermal process involving a non-ideal gas, we must return to Eq. (11.1) and use one of the following methods:

(1) direct graphical integration of a $P–v$ graph at constant temperature;

(2) substitution in Eq. (11.1) for the molar volume in terms of pressure and temperature, the equation of state of the real gas being known (see Example 1, page 210);

(3) as (2), but using the principle of corresponding states.

These methods are generally cumbersome in practice and, since the form of Eq. (11.3) is so simple, it would be preferable if we could keep the same form for a real gas. This is, in fact, what is normally done, a new function of state being defined, known as *fugacity, f,* which when substituted for pressure will make Eq. (11.2) hold for real gases at the temperature T:

$$d\mu = RT \, d \ln f \tag{11.8}$$

The fugacity of a gas is defined by this equation and the fact that real gases approach ideal behaviour as the pressure is decreased, i.e. as $P \rightarrow 0$, $f \rightarrow P$. Thus fugacity and pressure will be numerically equal at low pressures and deviate from each other only at high pressures when a real gas deviates from ideal behaviour. Fugacity may be regarded as a fictitious pressure or as an idealised measure of the pressure of a gas. We shall discuss later how it may be measured.

Integrating Eq. (11.8) at constant temperature between a standard state and the state of the gas considered, we have

$$\boxed{\mu - \mu^\ominus = RT \ln \frac{f}{f^\ominus}} \tag{11.9}$$

where $f^\ominus$ is the fugacity of the gas in the standard state.

Since most gases behave very nearly ideally at $101\ 325\ \text{N m}^{-1}$

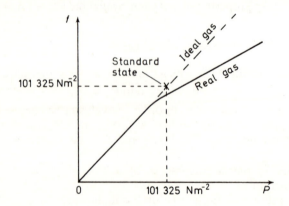

Figure 11.2. Hypothetical standard state

($= 1$ atm) pressure, we may, for most practical purposes, define *the standard state of a gas as the state of the pure gas at $101\ 325\ \text{N m}^{-2}$ pressure at the temperature considered.* However, strictly speaking, the definition of *the standard state of a gas is the state of the pure gas* *at a fugacity of* $101\ 325\ \text{N m}^{-2}$, *at the temperature under considera-tion if the gas behaved ideally* (see Figure 11.2).

This may seem a strange definition of a standard state but it is not particularly so, since the properties of ideal gases are very well known and thus the real gas may be compared with the ideal gas.

The ratio of the fugacity of a gas to its fugacity in the standard state, $f/f^{\ominus}$, is known as the *activity* of the gas,

$$\mu - \mu^{\ominus} = RT \ln a \qquad (11.10)$$

where a is the activity of the gas, $a = f/f^{\ominus}$. (11.11)

ALTERNATIVE APPROACH TO FUGACITY

For an ideal gas at constant temperature, Eq. (11.3) holds – that is,

$$\mu - \mu^{\ominus} = RT \ln \frac{P}{P^{\ominus}}$$

this being derived from

$$d\mu = RT\ d \ln P$$

Equation (11.3) obviously does not hold for non-ideal gases. How-ever, the form of this equation is so simple that we should like to have an equation of similar form for real gases. We shall therefore introduce a correction factor, ϕ, known as the fugacity coefficient, which is dependent on pressure and temperature, such that

$$d\mu = RT\ d \ln \phi P$$

holds for real gases at a given temperature. For ideal gases, $\phi = 1$.

The product

$$\phi P = f \qquad (11.12)$$

is known as the fugacity of the gas. Hence, integrating from the standard state to the state of the system being considered,

$$\mu - \mu^{\ominus} = RT \ln \frac{f}{f^{\ominus}}$$

The standard state is defined as previously. The fugacity coefficient is a measure of the deviation of the gas from ideal behaviour.

11.5   Determination of fugacity

(A)   FROM P, V, T DATA

For an ideal gas undergoing an isothermal process, we have, from Eqs (11.1) and (11.2),

$$RT\,\mathrm{d}\ln P = v_{\text{ideal}}\,\mathrm{d}P \tag{11.13}$$

where v_{ideal} is the molar volume of the ideal gas.

Similarly, for a real gas undergoing an isothermal process, we have, on combining Eqs (11.1) and (11.8),

$$RT\,\mathrm{d}\ln f = v\,\mathrm{d}P \tag{11.14}$$

where v is the molar volume of the real gas. Subtracting Eq. (11.13) from Eq. (11.14), we have

$$RT(\mathrm{d}\ln f - \mathrm{d}\ln P) = (v - v_{\text{ideal}})\,\mathrm{d}P$$

$$\mathrm{d}\ln\frac{f}{P} = \frac{1}{RT}(v - v_{\text{ideal}})\,\mathrm{d}P$$

Employing Eq. (11.12), and integrating at constant temperature from zero pressure to the pressure P, we have

$$\int_{1}^{\phi} \mathrm{d}\ln\phi = \frac{1}{RT}\int_{0}^{P}(v - v_{\text{ideal}})\,\mathrm{d}P$$

The limits on the left-hand side are 1 to ϕ since real gases approach ideal behaviour at low pressure, i.e. as $P \to 0$, the fugacity coefficient $\phi \to 1$, and at the pressure P the fugacity coefficient is ϕ. Hence,

$$\boxed{\ln\phi = \frac{1}{RT}\int_{0}^{P}(v - v_{\text{ideal}})\,\mathrm{d}P} \tag{11.15}$$

The right-hand side may be evaluated by graphical integration if $v - v_{\text{ideal}}$ is plotted against P.

Alternatively, use may be made of the fact that Eq. (11.15) may also be written as

$$\ln\phi = \int_{0}^{P}\left(\frac{v}{RT} - \frac{1}{P}\right)\mathrm{d}P \tag{11.16}$$

Once the fugacity coefficient has been found, the fugacity may be calculated using $f = \phi P$.

(B) FROM THE PRINCIPLE OF CORRESPONDING STATES

Equation (11.16) may be written slightly differently by making use of Eq. (11.7):

$$\ln \phi = \int_0^P (z-1) \frac{dP}{P} \tag{11.17}$$

where z is the compressibility factor.

Since $P_R = P/P_c$, $dP_R = dP/P_c$, so that $dP_R/P_R = dP/P$. Hence,

$$\ln \phi = \int_0^{P_R} \frac{(z-1)\,dP_R}{P_R} \tag{11.18}$$

By plotting $(z-1)/P_R$ against P_R at constant T_R, ϕ may be determined since $\ln \phi$ is equal to the area under the curve. Employing the principle of corresponding states, we may, by using the data in

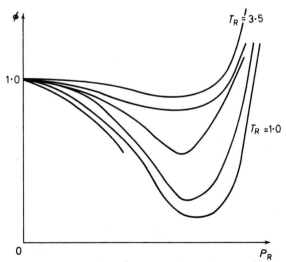

Figure 11.3. Graph of fugacity coefficient against reduced pressure for various reduced temperatures, Figure 11.1 being employed

Figure 11.1, draw a general fugacity coefficient chart which will give ϕ for all real gases (see Figure 11.3).

(C) FROM THE EQUATION OF STATE

From the definition of fugacity, Eq. (11.8), Eq. (11.1) may be written

$$\left(\frac{\partial \ln f}{\partial P}\right)_T = \frac{v}{RT} \tag{11.19}$$

By integrating this equation and then adjusting the integration constant so that $f \rightarrow P$ as $P \rightarrow 0$, we obtain an expression for f.

Example 1 The virial equation of state is

$$Pv = RT + BP + CP^2 + \cdots \tag{11.20}$$

Substituting for v in Eq. (11.19),

$$\left(\frac{\partial \ln f}{\partial P}\right)_T = \frac{1}{P} + \frac{B}{RT} + \frac{CP}{RT} + \cdots$$

Integrating at constant T,

$$\ln f = \ln P + \frac{BP}{RT} + \frac{CP^2}{2RT} + \cdots + \text{constant}$$

Now f tends to P as P tends to zero. Hence, the integration constant must be zero, giving

$$\ln f = \ln P + \frac{1}{RT}\left\{BP + \frac{CP^2}{2} + \cdots\right\} \tag{11.21}$$

or,

$$f = P \exp \frac{1}{RT}\left\{BP + \frac{CP^2}{2} + \cdots\right\}$$

Example 2 For a gas obeying van der Waals' equation, evaluate the fugacity as a function of volume.

Integrating Eq. (11.14) at constant T,

$$RT\int d \ln f = \int v \, dP$$

The variable on the right-hand side may be changed by integrating by parts, this making the calculation somewhat easier:

$$\int v \, dP = Pv - \int P \, dv$$

Also, van der Waals' equation may be written as

$$P = \frac{RT}{v-b} - \frac{a}{v^2}$$

Hence,

$$RT \ln f = Pv - \int \left(\frac{RT}{v-b} - \frac{a}{v^2} \right) dv$$

so that

$$RT \ln f = Pv - RT \ln (v-b) - \frac{a}{v} + \text{constant} \qquad (11.22)$$

The constant is evaluated by using the fact that $f \to P$ as $P \to 0$. Now as $P \to 0$ at constant T, the volume v becomes large, so that $v - b \to v$ and $1/v \to 0$. Hence,

$$\text{constant} = RT \ln RT - RT$$

Substituting this into Eq. (11.22),

$$RT \ln f = Pv - RT + RT \ln \frac{RT}{v-b} - \frac{a}{v}$$

van der Waals' equation may be rearranged as

$$Pv - RT = \frac{RTb}{v-b} - \frac{a}{v}$$

Hence,

$$\ln f = \ln \frac{RT}{v-b} + \frac{b}{v-b} - \frac{2a}{RTv} \qquad (11.23)$$

11.6 The effect of pressure on fugacity and activity

Differentiating Eq. (11.8) with respect to pressure at constant T,

$$RT \left(\frac{\partial \ln f}{\partial P} \right)_T = \left(\frac{\partial \mu}{\partial P} \right)_T$$

Employing Eq. (10.32) or Eq. (8.17) for a pure gas, we have:

$$\boxed{\left(\frac{\partial \ln f}{\partial P} \right)_T = \frac{v}{RT}} \qquad (11.24)$$

The effect of pressure on activity may easily be found, since the activity is proportional to the fugacity; $a = f/f^\circ$, and, hence,

$$\left(\frac{\partial \ln a}{\partial P}\right)_T = \frac{v}{RT}$$ (11.25)

11.7   The effect of temperature on fugacity and activity

Rather than begin with Eq. (11.8), it is easier to rearrange Eq. (11.9), which holds at a given temperature T, as

$$\ln f - \ln f^\circ = \frac{1}{R}\left\{\frac{\mu}{T} - \frac{\mu^\circ}{T}\right\}$$

where μ and μ° are measured at the same temperature. Differentiating with respect to T at constant P, we have

$$\left(\frac{\partial \ln f}{\partial T}\right)_P - \left(\frac{\partial \ln f^\circ}{\partial T}\right)_P = \frac{1}{R}\left(\frac{\partial(\mu/T)}{\partial T}\right)_P - \frac{1}{R}\left(\frac{\partial(\mu^\circ/T)}{\partial T}\right)_P$$

Employing Eq. (10.36), and remembering that f° is chosen such that the gas behaves ideally at 101 325 N m^{-2} at all temperatures, i.e.

$$\left(\frac{\partial \ln f^\circ}{\partial T}\right)_P = 0$$

we obtain:

$$\left(\frac{\partial \ln f}{\partial T}\right)_P = -\frac{\bar{H} - \bar{H}^\circ}{RT^2}$$ (11.26)

where $\bar{H}$ and $\bar{H}^\circ$ are the partial molar enthalpies of the pure gas (which are, of course, the molar enthalpies) at temperature T in the state at pressure P and in the standard state where the gas behaves ideally (i.e. at low pressure), respectively.

Equation (11.26) may be written as

$$\left(\frac{\partial \ln f}{\partial T}\right)_P = -\frac{h - h^\circ}{RT^2}$$ (11.27)

The pressure dependence for activity is identical with this equation.

11.8   The fugacity of liquids and solids

Unlike gases, there are no general equations of state for liquids or solids. Hence, it is necessary to determine their properties by the graphical integration of P, V, T data or from different experimental equations of state for each substance. It would be highly convenient (and certainly rather elegant) if we could use an equation of the same form as Eq. (11.3) whether we were dealing with a gas, a liquid or a solid. Hence, for a substance i at constant T, we write

$$d\mu_i = RT \, d \ln f_i \qquad (11.28)$$

where f_i is the fugacity of the liquid (or solid).

Integrating from a standard state to the state of the system being considered,

$$\mu_i = \mu_i^\ominus + RT \ln \frac{f_i}{f_i^\ominus} \qquad (11.29)$$

All liquids and solids have a definite vapour pressure at a given temperature. Consider a liquid substance i in equilibrium with its vapour. The chemical potential of i in the liquid phase is equal to that in the vapour phase,

$$\mu_i^{liq} = \mu_i^{vap}$$

Now consider an isothermal process (e.g. the introduction of an inert gas in order to increase the pressure). Since before and after the process the liquid and vapour are in equilibrium,

$$d\mu_i^{liq} = d\mu_i^{vap}$$

Employing Eq. (11.28) for both phases, we have

$$RT \, d \ln f_i^{liq} = RT \, d \ln f_i^{vap}$$

At low pressure, fugacity is equal to pressure. Hence, on integrating, the constant of integration is zero and

$$\boxed{f_i^{liq} = f_i^{vap}} \qquad (11.30)$$

That is, the fugacity of a substance i in any phase (liquid or solid) is equal to the fugacity of i in the vapour phase when the vapour and other phase are in equilibrium.

Alternatively, since the liquid and vapour are in equilibrium, the force exerted on a liquid molecule near the surface must be the same as that exerted on a vapour molecule. Since pressure is defined as force per unit area,

$$P_i^{liq} = P_i^{vap}$$

(assuming ideal behaviour), or,

$$f_i^{liq} = f_i^{vap}$$

The vapour pressures of ice and water are equal at the triple point. This is true for any substance i existing in a number of phases that are in equilibrium, the vapour pressure of i for each phase being the same. Therefore, whenever a number of phases of a substance are in equilibrium, the fugacity of the substance is the same in each of the phases.

Since, in general, the vapour pressure of solids is low, *the fugacity of a solid is equal to its vapour pressure.* Similarly, *for liquids, the fugacity is approximately equal to the vapour pressure.*

The fugacity of a pure liquid or solid at *any* temperature and pressure may be found from the experimental value of the fugacity of the vapour at the saturation temperature and pressure, T_{sat} and P_{sat} (the saturation state being that when the liquid or solid phase is in equilibrium with its vapour). By using Eq. (11.24), we may calculate the fugacity at some other pressure P at T_{sat}, or, by employing Eq. (11.27), we may calculate the fugacity at some other temperature T at P_{sat}.

Example   Calculate the fugacity of water at 298 K and 10^7 N m^{-2}, given that at 298 K the vapour pressure of water is 3160 N m^{-2} and the density of water is 997·04 kg m^{-3}, the molecular weight being 18·02.

We have that the molar volume is given by

$$v = \frac{18 \cdot 02}{997 \cdot 04 \times 10^3} = 1 \cdot 807 \times 10^{-5} \text{ m}^3 \text{ mol}^{-1}$$

Since the vapour pressure is only 3160 N m^{-2} at 298 K, we may take this as being the fugacity of the liquid at 298 K. Integrating Eq. (11.24) at constant T from state 1 to state 2, and remembering that, in general, the volume of a liquid does not vary much with pressure,

$$\boxed{RT \ln \frac{f_2}{f_1} = v(P_2 - P_1)} \tag{11.31}$$

Therefore $RT \ln \dfrac{f}{3160} = 1 \cdot 807 \times 10^{-5} (10^7 - 3160)$

$$f = 3402 \text{ N m}^{-2}$$

STANDARD STATES FOR LIQUIDS AND SOLIDS

Although the value of fugacity is independent of the standard state, since it is in fact equal to the pressure at low pressures, activity,

$a_i = f_i/f_i^\circ$, is not. Since the fugacity of a substance in the liquid or solid phase is usually small, it is convenient to use a different standard state from that for the substance in the gaseous phase.

The standard state for liquids is the state of the pure liquid at its vapour pressure and at the temperature under consideration.

The standard state for solids is defined similarly.

Another standard state that is often used for liquids (or solids) is *the state of the pure liquid (or solid) at a total pressure of 101 325 N m^{-2} and at the temperature under consideration.*

However, we have seen from the example on page 203 that, in general, the vapour pressure of a liquid does not alter very much with increase in the total pressure. We shall therefore consider the two types of standard state as being the same.

The activity is seen to be equal to unity in the standard state. From the example on page 214, it is seen that the fugacity of water in the standard state (pure water at 298 K and 3160 N m^{-2}) is 3160 N m^{-2}. Hence, the activity of water at 298 K and 10^7 N m^{-2} is given by

$$a = \frac{f}{f^\circ} = \frac{3402}{3160} = 1 \cdot 077$$

At relatively low pressures, less than ~ 1 MN m^{-2} (~ 10 atm), it is seen from Eq. (11.31) and the example above that the change in the fugacity of a liquid or solid with the total pressure, at constant temperature, is negligible. This is equivalent to saying that the vapour pressure of a liquid or solid does not alter drastically with pressure (see example on page 203). Hence, *at low pressures, the activity of a pure liquid or solid may be considered to be unity.*

We have seen that the chemical potential of a liquid (or solid) may be given by

$$\mu_i = \mu_i^\circ + RT \ln \frac{f_i}{f_i^\circ}$$

We have also seen that the vapour pressure of a liquid (or solid) does not, in general, vary greatly with the total pressure. Thus, if the pressure is not high, $f_i \doteqdot f_i^\circ$ and, hence,

$$\boxed{\mu_i = \mu_i^\circ} \tag{11.32}$$

is a very good approximation at pressures less than about 1 MN m^{-2} (~ 10 atm).

Problems

1. Show that for a reversible adiabatic change of a gas obeying van der Waals' equation, $P(v-b)^{\gamma-1} = $ constant.

2.   Show that for a gas obeying the equation of state $P(v-b) = RT$, where b is temperature dependent, the fugacity is $f = P \exp(bP/RT)$.

3.   Calculate the fugacity of a real gas at 5 MN m^{-2} if at a given temperature the following results for the compressibility are obtained:

$$P/\text{MN m}^{-2} \quad 0 \quad 1\cdot42 \quad 2\cdot23 \quad 3\cdot04 \quad 4\cdot45 \quad 5\cdot87$$
$$z \qquad\qquad 1 \quad 0\cdot941 \quad 0\cdot921 \quad 0\cdot892 \quad 0\cdot841 \quad 0\cdot791$$

Answer $4\cdot24 \text{ MN m}^{-2}$

4.   The vapour pressure of mercury is $36\cdot4 \text{ N m}^{-2}$ at a temperature of $373\cdot2 \text{ K}$ and pressure 10^5 N m^{-2}. If the density of mercury is $1\cdot335 \times 10^4 \text{ kg m}^{-3}$, calculate the vapour pressure at $373\cdot2 \text{ K}$ and a pressure of 10^8 N m^{-2}.

Answer $60\cdot66 \text{ N m}^{-2}$

5.   The vapour pressure of a liquid is $33\cdot33 \text{ N m}^{-2}$ at a pressure of 10^5 N m^{-2} and at 293 K. If its density is $4\cdot93 \times 10^3 \text{ kg m}^{-3}$, calculate its vapour pressure at 10^8 N m^{-2}.

Answer $280\cdot0 \text{ N m}^{-2}$

12
Phase Equilibria

12.1 The P, V, T surface

Phase equilibria are independent of the actual amounts of each phase present. For example, liquid water and ice exist in equilibrium at $101\,325\ N\ m^{-2}$ and $273{\cdot}15\ K$, it being immaterial how much liquid or ice is present.

We shall first discuss pure substances, i.e. one-component systems, in the absence of external fields. We have seen previously that the

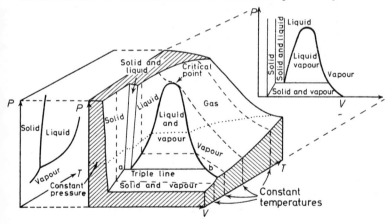

Figure 12.1. Surface formed on a $P-V-T$ diagram by the equilibrium states of a one-component system which contracts on freezing. The $P-T$ and $P-V$ projections of the surface are shown

state of a closed system consisting of a pure homogeneous substance of given mass may be described by any two independent functions of state, e.g. P, V or T, V, etc. If we plot three functions of state along three mutually perpendicular axes for a given substance, the result is a surface which represents all possible *equilibrium states* of the

217

substance. We would *normally* choose *P*, *V* and *T*, since these properties are relatively easy to measure.

A typical surface is shown in Figure 12.1 for a substance which contracts on freezing. The projections of the surface on the *P–T* and *P–V* planes are shown to the left and right, respectively. In Figure 12.2 is shown a typical surface for a substance which expands on

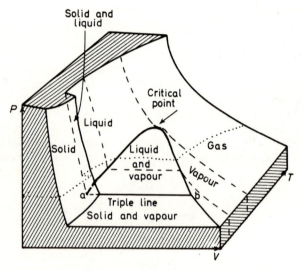

Figure 12.2. The P–V–T surface of equilibrium states for a system which expands on freezing

freezing (e.g. water). The diagrams are not to scale, the volume axes having been considerably foreshortened.

In those regions of the surface labelled solid, liquid and gas, only one phase may exist; in those regions labelled solid–vapour, solid–liquid and liquid–vapour, two phases may exist in equilibrium. In the regions where two phases may exist, lines may be ruled along the surface parallel to the *V*-axis. These represent phase changes (melting, vaporisation or sublimation), which take place at constant temperature. (Sublimation is the direct transformation of solid to vapour.) Along the triple line, *ab*, which is a line at constant *P* and *T*, all three phases may exist in equilibrium.

Rather than discuss the surface in greater detail, we shall turn our attention to the *P–T* and *P–V* projections. These and similar types of projections employing any other functions of state are known as *phase diagrams.*

12.2 The P–V diagram

A saturation state is any state represented by a point on a line separating a one-phase from a two-phase region. Hence, the curve cd is known as the saturation–liquid curve, and bc as the saturation–vapour curve.

The two-phase region bcd, labelled liquid and vapour, is known as the *wet region*. A state given by a point inside this region consists

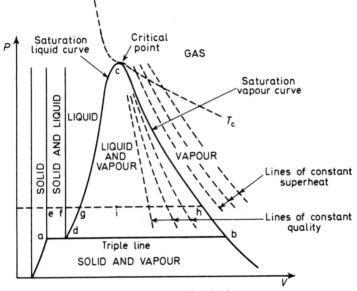

Figure 12.3. P–V projection

of saturated liquid and saturated vapour. The proportion of vapour and liquid in the system is denoted by the *quality*:

> quality = *fraction of the system which is vapour, by mass*

Hence, the quality of a system consisting of a saturated liquid, i.e. a state on the line cd, is zero, and the quality of a system consisting of a saturated vapour, i.e. a state on the line bc, is unity. Some lines of constant quality are shown in Figure 12.3. If the state of the system is given by a point i, then the quality = gi/gh.

The substance is said to be in a superheated state if its temperature at a given pressure is greater than the liquid–vapour saturation temperature, i.e. states to the right of the line bc.

Degree of superheat equals the temperature in excess of the liquid–vapour saturation temperature at a given pressure.

The volume change during a solid–liquid phase change is very small and equal to ef at a given pressure, e.g. when 1 mole of ice is converted to liquid water at 273·15 K, the volume change is $1·7326 \times 10^{-6}$ m^3. However, the volume change during a liquid–vapour phase change is very large and equal to gh at the same pressure, e.g. when 1 mole of liquid water is converted to steam at 373·15 K, the volume change is $3·0078 \ 10^{-2}$ m^3. This illustrates how greatly the V-axis has been foreshortened (db is several thousand times greater than ad).

However, it is seen from Figure 12.3 that the volume change during a liquid–vapour phase change at higher pressures, and hence higher temperatures, decreases and becomes zero at the point c. Above this *critical point*, the process of vaporisation cannot take place. We may define the critical temperature T_c of a substance as being the temperature above which it is impossible for condensation to occur no matter how great a pressure is applied. Or we may say that above the critical state, i.e. at pressures and temperatures greater than P_c and T_c, we cannot distinguish between the liquid and gas phases. It is usual to refer to a substance as being a gas if its temperature is greater than T_c.

Below the triple line, liquid cannot exist.

The T–V diagram is similar to that of the P–V.

12.3 The P–T diagram

The projection of the surface in Figure 12.1 on the P–T plane gives Figure 12.4, e.g. the projection of the liquid–vapour phase surface is the vaporisation curve. A system which has a state represented by a point on one of the curves shown consists of two phases in equilibrium, e.g. a point on the vaporisation curve represents the state of a system consisting of liquid and vapour in equilibrium. Hence, the vaporisation curve shows how the temperature at which the liquid boils (or the vapour condenses) varies with pressure. Similarly, the fusion curve indicates the variation with pressure of the temperature at which the solid melts, and the sublimation curve shows the pressure dependence of the sublimation temperature. Boiling point and melting point, strictly speaking, refer to the temperature at which the liquid boils or the solid melts at 101 325 N m^{-2}.

At the triple point all three phases – liquid, solid and vapour – exist in equilibrium. States above the vaporisation curve have pressures greater than the pressure of the saturation pressure at a given temperature, and are thus sometimes referred to as *compressed liquid* states. When dealing with the vapour, the saturation

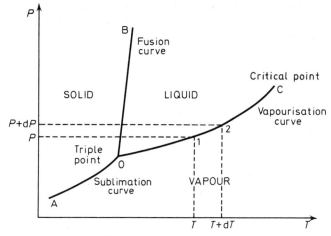

Figure 12.4. P–T projection

pressure is sometimes known as the saturation vapour pressure, s.v.p. Similarly, if the temperature of the substance is greater than that of the liquid–vapour saturation temperature at a given pressure, the substance is said to be in a *superheated state*, which has been mentioned in connection with the *P–V* diagram.

12.4 The Clapeyron equation

The Clapeyron equation gives the dependence of the pressure on temperature when two phases of a pure substance are in equilibrium. In other words, it gives the dependence of the saturation pressure on the saturation temperature.

A typical phase diagram for a pure substance is given in Figure 12.4. Consider an equilibrium system consisting of two phases, α and β, of a pure substance. For example, let the system have state 1 on line OC. We have shown previously that for phase equilibrium the chemical potential of a component i in phase α is equal to that of the component i in phase β

$$\mu_i^\alpha = \mu_i^\beta$$

Since the substance is pure, the chemical potentials are identical to the molar Gibbs function; therefore

$$g^\alpha = g^\beta \tag{12.1}$$

Now consider another slightly different equilibrium state (state 2) for which the Gibbs functions per mole for the two phases are $g^\alpha + dg^\alpha$ and $g^\beta + dg^\beta$ at the pressure $P + dP$ and temperature $T + dT$. Since the two phases are in equilibrium,

$$g^\alpha + dg^\alpha = g^\beta + dg^\beta \tag{12.2}$$

Subtracting Eq. (12.1) from Eq. (12.2)

$$dg^\alpha = dg^\beta$$

Substituting for dg from Eq. (8.15) for each phase,

$$v^\alpha \, dP - s^\alpha \, dT = v^\beta \, dP - s^\beta \, dT \tag{12.3}$$

where v^α and v^β are the molar volumes of phases α and β and s^α and s^β are the molar entropies of phases α and β, all at pressure P and temperature T.

Rearranging Eq. (12.3),

$$(v^\alpha - v^\beta) \, dP = (s^\alpha - s^\beta) \, dT$$

$$\frac{dP}{dT} = \frac{s^\alpha - s^\beta}{v^\alpha - v^\beta} = \frac{\Delta s}{\Delta v} \tag{12.4}$$

From Eq. (5.45), the entropy change during a phase change is given by

$$\Delta s = \frac{\Delta h}{T}$$

where T is the temperature at which the phase change occurs.

Hence,
$$\boxed{\frac{dP}{dT} = \frac{\Delta h}{T \Delta v}} \tag{12.5}$$

This equation is known as the *Clapeyron equation* and is applicable to any phase change–fusion, vaporisation, sublimation, or changes between crystalline forms, etc. It is also sometimes referred to as the Clausius–Clapeyron equation although it was, in fact, derived by Clapeyron.

The Clapeyron equation enables the evaluation of enthalpy

changes for phase changes from a knowledge of vapour pressure data only.

12.5   The Clausius–Clapeyron equation

There are several approximations to the Clapeyron equation, the best-known being applicable to the vapour–liquid and vapour–solid equilibria of a pure substance:

$$\frac{\mathrm{d}P}{\mathrm{d}T} = \frac{\Delta h}{T(v_{\mathrm{vap}} - v_{\mathrm{liq}\,(\text{or solid})})}$$

where Δh is the molar latent heat of vaporisation (or sublimation), v_{vap} is the molar volume of the vapour, and $v_{\mathrm{liq}\,(\text{or solid})}$ is the molar volume of the liquid (or solid).

The molar volume of the vapour is considered to be very much larger than that of the liquid (or solid). Hence, $v_{\mathrm{liq}\,(\text{or solid})}$ *is negligible compared with* v_{vap}:

$$\frac{\mathrm{d}P}{\mathrm{d}T} = \frac{\Delta h}{T v_{\mathrm{vap}}}$$

If we make the further assumption that the *vapour behaves as an ideal gas*, then the molar volume is given by $v_{\mathrm{vap}} = RT/P$, giving

$$\frac{\mathrm{d}P}{\mathrm{d}T} = \frac{P\Delta h}{RT^2} \tag{12.6}$$

Since $\mathrm{d}P/P = \mathrm{d}\ln P$, we therefore have

$$\frac{\mathrm{d}\ln P}{\mathrm{d}T} = \frac{\Delta h}{RT^2} \tag{12.7}$$

This equation is known as the *Clausius–Clapeyron equation*.

If we assume that Δh is *constant* between state 1 and state 2, we can integrate this equation,

$$\int_1^2 \mathrm{d}\ln P = \frac{\Delta h}{R} \int_1^2 \frac{\mathrm{d}T}{T^2}$$

$$\ln \frac{P_2}{P_1} = -\frac{\Delta h}{R}\left(\frac{1}{T_2} - \frac{1}{T_1}\right) \tag{12.8}$$

Integrating without limits,

$$\ln P = -\frac{\Delta h}{RT} + \text{constant} \qquad (12.9)$$

If the logarithm of the *saturated vapour pressure*, ln *P*, is plotted against the reciprocal of the absolute temperature, a straight line with slope $-\Delta h/R$ is obtained. Hence, using the results from vaporisa-

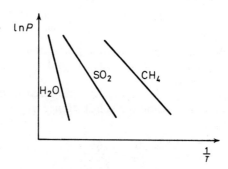

Figure 12.5. Graph of ln *P against* $1/T$ *for various gases*

tion or sublimation curves, a series of lines as in Figure 12.5 may be obtained.

If, in Eq. (12.9), we let the constant equal ln *A*, then this equation may be written as

$$P = A \exp\left(-\frac{\Delta h}{RT}\right) \qquad (12.10)$$

Example Calculate the latent heat of vaporisation of water if $dP/dT = 361\cdot4\ \text{N m}^{-2}\ \text{K}^{-1}$ and the molar volumes of water and steam are $18\cdot73 \times 10^{-6}$ and $30\,130 \times 10^{-6}\ \text{m}^3\ \text{mol}^{-1}$. The boiling point of water is 373·15 K.

From Eq. (12.5), the latent heat of vaporisation of water, Δh_v, is given by

$$\Delta h_v = 373\cdot15(30\,130 - 18\cdot73) \times 10^{-6} \times 361\cdot4$$
$$= 40\cdot61\ \text{kJ mol}^{-1}$$

12.6 Trouton's rule

An approximate value for a latent heat may be found from Trouton's rule, which states that

$$\frac{\Delta h_v}{T_B} \doteq 88 \text{ J K}^{-1} \text{ mol}^{-1} \tag{12.11}$$

where Δh_v = molar latent heat of vaporisation of a substance, and T_B = boiling point of the substance (pressure = $101\,325 \text{ N m}^{-2}$): That is, the entropy of vaporisation for all substances is, approximately, constant ($88 \text{ J K}^{-1} \text{ mol}^{-1}$). Trouton's rule holds well for non-polar liquids of molecular weight about 100. If hydrogen-bonding occurs in the liquid, e.g. ethanol or water, the ratio is greater than that predicted, since the latent heat includes energy required to break the forces between the molecules of the liquid.

From Eq. (12.8), using the boiling point and an approximate value for the latent heat, the vapour pressure at some other temperature may be calculated:

$$\ln\frac{P}{101\,325} = -\frac{\Delta h_v}{R}\left(\frac{1}{T}-\frac{1}{T_B}\right)$$

Of course, it should be noted here that the latent heat of vaporisation does, in fact, vary with temperature and pressure, and therefore the integration which resulted in Eq. (12.8) is not justified. Strictly, the temperature dependence of Δh must be taken into account.

Example   The boiling point of benzene is 353 K. Assuming Trouton's rule is obeyed, calculate the vapour pressure at 303 K. ($\ln N = 2\cdot303 \log_{10}N$)

From Trouton's rule, the latent heat of vaporisation is given by

$$\Delta h_v = 88 \times 353 \text{ J mol}^{-1}$$
$$= 31\cdot07 \text{ kJ mol}^{-1}$$

The vapour pressure at the boiling point must be $101\,325 \text{ N m}^{-2}$ (1 atm). Hence, from Eq. (12.8),

$$\ln\frac{P}{101\,325} = -\frac{31\,070}{R}\left(\frac{1}{303}-\frac{1}{353}\right)$$

$$P = 17\,650 \text{ N m}^{-2}$$

12.7 The T–S diagram

The *T–S* diagram for a pure substance is shown in Figure 12.6, the dotted lines representing isobars. It must be remembered that the area under a curve on a $T\text{-}S$ diagram is equal to the heat absorbed during a reversible process from the initial to the final state. Hence, the area under AB is equal to the latent heat of fusion at the pressure given by the isobar, and the area under CD is equal to the latent heat of vaporisation at the same pressure. Therefore it can be seen that, at the triple point, the sum of the latent heat of fusion, L_f, and

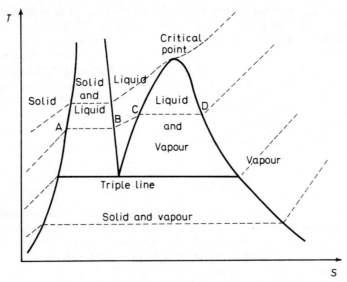

Figure 12.6. *T–S diagram for a one-component system*

the latent heat of vaporisation, L_v, must be equal to the latent heat of sublimation, L_s. At the triple point,

$$L_s = L_f + L_v \qquad (12.12)$$

It can also be seen from the diagram that the latent heat of vaporisation decreases with increase in temperature and becomes zero at the critical point.

*12.8   Calculation of quality

In Figure 12.7 pressure P is plotted against specific volume v. It can be seen that the specific volume of the vapour, v_g, is the sum of the specific volume of the liquid, v_f, and the specific volume change from liquid to vapour, v_{fg},

$$v_g = v_f + v_{fg}$$

The specific volume, v, of a system in state C is the sum of the volumes of the liquid and the vapour in the system,

$$v = (1-x)v_f + xv_g$$

where x is the quality, i.e. the fraction by mass of the system which is vapour. Rearranging,

$$v = v_f + xv_{fg} \qquad (12.13)$$

Similar expressions hold for enthalpy and entropy:

$$h = h_f + xh_{fg} \qquad (12.14)$$
$$s = s_f + xs_{fg} \qquad (12.15)$$

h_{fg} is the latent heat of vaporisation and s_{fg} is the specific entropy of vaporisation,

$$\boxed{s_{fg} = \frac{h_{fg}}{T} = \frac{\Delta h_v}{T}} \qquad (12.16)$$

where T is the temperature of the system in state C (Figure 12.8).
The specific entropy of the system in state A is given by

$$s_f = s_0 + \int_{T_0}^{T} \frac{c_P(\text{sat. liq.})\, dT}{T}$$

where s_0 is the specific entropy of the substance in a reference state

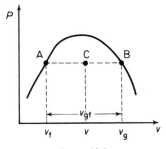

Figure 12.7.

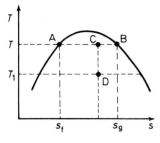

Figure 12.8.

of temperature T_0. Assuming c_P to be temperature independent,

$$s_f = s_0 + c_P \ln \frac{T}{T_0} \qquad (12.17)$$

In steam tables it is generally taken that $T_0 = 273 \cdot 15\,\text{K}$ and the enthalpy and entropy of the saturated liquid at this temperature are zero.

For a reversible adiabatic (and, hence, isentropic) change, $s_C = s_D$. Therefore, substituting into Eq. (12.15), using Eqs (12.16) and (12.17), we have

$$c_P \ln \frac{T}{T_0} + x_C \frac{\Delta h_C}{T} = c_P \ln \frac{T_1}{T_0} + x_D \frac{\Delta h_D}{T_1}$$

or,

$$c_P \ln T + x_C \frac{\Delta h_C}{T} = c_P \ln T_1 + x_D \frac{\Delta h_D}{T_1} \qquad (12.18)$$

where Δh_C and Δh_D are the latent heats of vaporisation at temperatures T and T_1. Hence, x_D may be calculated from a knowledge of the other parameters in the equation.

12.9   Second-order transitions

During ordinary phase transitions, changes in volume, entropy and enthalpy (latent heat) are observed. The heat capacity during a transition is infinite, since a finite quantity of heat (latent heat) is required to raise the temperature infinitesimally.

In Figure 12.9, the effects of a typical phase change on μ, V, S, C_P, $v\kappa$, are shown. The chemical potential changes gradient, but is

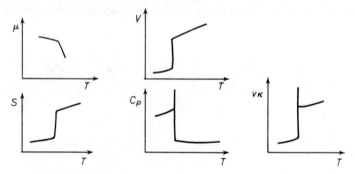

Figure 12.9. The effect of a first-order phase change on various thermodynamic properties

not discontinuous as are V, S, C_P. This is because, during a phase change, the chemical potential of a component is the same in both phases (see Eq. 10.29). The enthalpy change during the phase change (i.e. the latent heat) is given by $T\Delta S$.

For a pure substance, the first and second derivatives of μ with respect to T and P are

$$\left(\frac{\partial \mu}{\partial T}\right)_P = -s, \quad \left(\frac{\partial \mu}{\partial P}\right)_T = v$$

$$\left(\frac{\partial^2 \mu}{\partial T^2}\right)_P = -\left(\frac{\partial s}{\partial T}\right)_P = -\frac{c_P}{T}, \quad \left(\frac{\partial^2 \mu}{\partial P^2}\right)_T = \left(\frac{\partial v}{\partial P}\right)_T = -v\kappa$$

where κ is the isothermal compressibility.

The usual types of phase changes are known as *first-order transitions* since the first derivatives of μ are discontinuous at the phase transition.

The effect of a *second-order transition* on μ, V, S, C_P is shown in Figure 12.10. In this unusual type of phase change the second derivative of μ is discontinuous. The characteristic shape of the C_P curve resembles the Greek letter Λ; hence, the name *lambda transition*

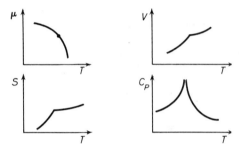

Figure 12.10. The effect of a second-order phase change on various thermodynamic properties

is often used. The enthalpy difference between the two phases is zero or almost zero. The heat capacity alters rapidly as the transition point is approached. This indicates that the phase change starts even before the transition temperature is reached.

Certain alloys show this behaviour, a typical example being brass (Zn:Cu as 1:1). The crystalline structure is body-centred cubic (Figure 12.11). At normal temperatures each Zn atom is surrounded by eight Cu atoms, each Cu atom being similarly surrounded by eight Zn atoms (Figure 12.12). The energy of interaction between similar atoms is less than that between Zn and Cu atoms.

As the temperature is increased, more and more Zn and Cu atoms have enough energy to exchange places. At the λ-transition point a random arrangement of Zn and Cu atoms is achieved. The heat absorbed per degree temperature rise (i.e. C_P) is greatest at the λ-point, which indicates that the greatest amount of disorder occurs at this point. Beyond the λ-point the heat capacity drops to the value expected.

Certain polymers and chemical compounds exhibit second-order transitions due to similar order–disorder transitions. Symmetrical ions in the crystal lattice of a salt do not rotate freely at low tempera-

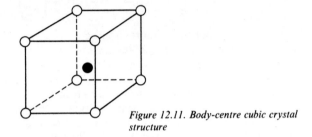

Figure 12.11. *Body-centre cubic crystal structure*

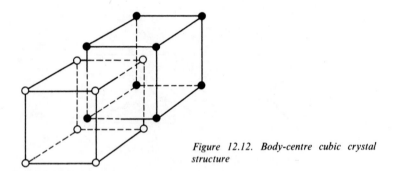

Figure 12.12. *Body-centre cubic crystal structure*

tures and a λ-transition occurs owing to the change from small oscillations to free rotation. Other examples are the transition from helium I to helium II, the transition of a substance from a ferromagnetic to a para-magnetic state at the Curie point and the transition of a metal from a superconducting state to normal conduction in the absence of a magnetic field.

12.10   The phase rule

Consider a closed system containing $\mathscr{P}$ phases in contact in which external fields (e.g. magnetic) and surface-tension effects are absent. Let the system contain C non-reacting components.

We shall now calculate *the number of degrees of freedom F of the system*, which may be defined as *the minimum number of intensive properties required to fix the values of all the remaining intensive properties of the system at equilibrium.*

The state of a phase may be given by the temperature, pressure and amount of each component in the phase. The composition of the phase may be given by $C-1$ mole fractions or weight percentages. Hence, the state of a phase is specified by $C+1$ intensive properties, including T and P (as confirmed in Section 10.3).

If all the phases $\mathscr{P}$ of a system are separated from one another by rigid boundaries, then *the state of the system is specified by $\mathscr{P}(C+1)$ intensive properties* (Figure 12.13).

Suppose that when (the rigid boundaries being removed) the phases are in contact they are in equilibrium. Then there exist relations which reduce the number of independent variables of the system. If the system is in *thermal equilibrium*, the temperature of all the phases must be the same,

$$T_A = T_B = \cdots = T_{\mathscr{P}}$$

Therefore we have overspecified the temperature $\mathscr{P}-1$ times.

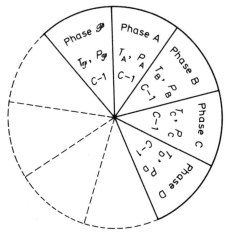

Figure 12.13. Phases A, B, ... $\mathscr{P}$ in separate compartments

If the system is in *mechanical equilibrium*, the pressure of each phase must be the same,

$$P_A = P_B = \cdots = P_{\mathscr{P}}$$

so that we have overspecified the pressure $\mathscr{P}-1$ times. If the system is in *chemical equilibrium*, the chemical potential of a component must be the same in all phases, otherwise there would be a net flow of that component between the phases. Hence,

for component 1, $\mu_1^A = \mu_1^B = \cdots = \mu_1^{\mathscr{P}}$ – there are $\mathscr{P}-1$ restrictions
for component 2, $\mu_2^A = \mu_2^B = \cdots = \mu_2^{\mathscr{P}}$ – there are $\mathscr{P}-1$ restrictions

for component C, $\mu_C^A = \mu_C^B = \cdots = \mu_C^{\mathscr{P}}$ – there are $\mathscr{P}-1$ restrictions

Therefore the total number of restrictions including those on T and P

$$= (C+2)(\mathscr{P}-1)$$

Hence, the number of degrees of freedom,

$$F = \mathscr{P}(C+1) - (C+2)(\mathscr{P}-1),$$

i.e.

$$\boxed{F + \mathscr{P} = C + 2} \tag{12.19}$$

This is the well-known *phase rule*, which was first derived by Gibbs.

For systems in which some of the substances present may react chemically, the *number of components, C*, is defined as *the minimum number of independently variable chemical species necessary to describe the composition of all the phases in the system.*

We may, in general, say that the number of components is equal to the number of chemical species in the system minus the number of independent chemical equilibria present. For example, the three-phase system containing $CaCO_3$, CaO, CO_2 has three chemical species present. However, since we have the reaction $CaCO_3 = CaO + CO_2$, there are $3 - 1 = 2$ components. That is, if the concentrations of two of the chemical species, e.g. CaO and CO_2, are known, then the third concentration may be found by using the equilibrium constant.

Illustration of the phase rule For *any* one-component system, i.e. $C = 1$, $F + \mathscr{P} = 3$,

if the system consists of one phase, $\mathscr{P} = 1$, then $F = 2$

if the system consists of two phases, $\mathscr{P} = 2$, then $F = 1$

if the system consists of three phases, $\mathscr{P} = 3$, then $F = 0$

For simplicity, consider Figure 12.4. In order to specify the intensive state of a system in a one-phase region, we require two variables, in this case temperature and pressure. To specify the intensive state of a two-phase system, we require only one variable, e.g. the state 1 is given by either temperature or pressure, since there is only one constant temperature or pressure line which cuts the curve OC at the point 1.

At the triple point, O, there are three phases and thus no degrees of freedom. There is only one particular pressure and temperature at which solid, liquid and vapour exist in equilibrium. Hence, if one variable is slightly altered, then one or two phases will disappear. In other words, it is sufficient to say that a system consists of ice, steam and liquid water to define the intensive state of the system.

12.11 Binary solutions

A solution is any homogeneous mixture of two or more chemical species. The phase of the solution may be gaseous, liquid or solid – for example, an alloy is a solid solution. Although gases may be mixed in all proportions, most liquids and solids are not miscible in all proportions and therefore their behaviour is much more complicated.

For any two-component (binary) system, $C = 2$, so that $F + \mathcal{P} = 4$. Hence, if the system consists of one phase, $\mathcal{P} = 1$, so that $F = 3$. That is, we require three variables to define the intensive state of a one-phase system. These three variables, normally temperature, pressure and the mole fraction of one of the compounds, may be plotted on a three-dimensional diagram as in Section 12.1. On such a diagram the equilibrium states of pure substances appear as surfaces, but for binary systems the equilibrium states fill the entire three dimensions. Just as before, for simplicity a two-dimensional diagram is preferable, and therefore either a constant-temperature or a constant-pressure plane is considered.

12.12 Raoult's law

Just as the concept of an ideal gas is useful, mainly as a means of comparison for real gases, the concept of an *ideal solution* is useful in the study of solutions. Ideality of a gas implies the complete absence of cohesive forces. Ideality of a solution implies complete uniformity of cohesive forces. For a two-component solution of A and B, the forces between the molecules A and A, A and B, B and B are equal.

The partial vapour pressure of a component above a solution may be regarded as a measure of that component's tendency to escape from the solution. (Strictly speaking, this is only so if the vapour behaves ideally, i.e. there are no attractive forces in the vapour.) Consider a solution of A and B with mole fractions x_A and x_B. For an ideal solution, the tendency of a molecule A to escape into the vapour phase is the same whether the A molecule is entirely surrounded by A or B molecules or partly by A and partly

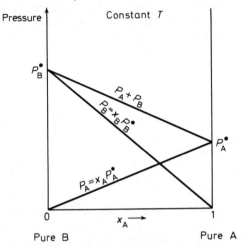

Figure 12.14. Raoult's law

by B. In other words, each A molecule behaves as if it were in pure A liquid. Hence, the partial vapour pressure, P_A, of component A is proportional to the relative amount of A in the solution.

For an ideal solution, the vapour pressure of each component is proportional to the mole fraction of that component in the solution.

If the solution remains ideal as the mole fraction x_A of component A approaches unity, the partial vapour pressure P_A must approach the vapour pressure $P_A^\bullet$ of the pure liquid A at that temperature. Hence,

$$P_A = x_A P_A^\bullet \qquad (12.20)$$

This equation is a statement of Raoult's law. For a solution obeying Raoult's law, all the components obey equations of the form $P_i = x_i P_i^\bullet$. It should be noted that Raoult's law is restricted to constant temperature and constant total pressure. For a binary solution, the total vapour pressure is $P = P_A + P_B = x_A P_A^\bullet + x_B P_B^\bullet$.

A solution is said to be ideal if each component obeys Raoult's law. A solution whose components obey Raoult's law *over all compositions* at a given temperature is known as a *perfect solution*.

Mixtures of components which are chemically similar generally form perfect solutions, e.g. benzene–toluene, n-heptane–n-hexane, ethylene bromide–propylene bromide.

For a perfect two-component solution, the total vapour pressure above the solution is given by

$$P = x_A P_A^\bullet + x_B P_B^\bullet = x_A P_A^\bullet + (1 - x_A) P_B^\bullet$$

i.e.
$$P = x_A (P_A^\bullet - P_B^\bullet) + P_B^\bullet$$

Therefore the total vapour pressure is a linear function of the mole fraction x_A, as seen in Figure 12.14. This figure shows the variation of the partial vapour pressures of both components A and B of a binary perfect solution and the total vapour pressure of the solution with the composition of the solution at constant temperature.

We shall show later that for, ideal solutions, the enthalpy and volume changes on mixing the pure components are zero.

12.13 Solutions exhibiting deviations from Raoult's law

Solutions for which the vapour pressures of the components are less than those predicted by Raoult's law are said to exhibit *negative deviations*, as in Figure 12.15. On this graph the vapour pressures of A and B and the total pressure are plotted against mole fraction. The dotted lines show the values predicted by Raoult's law. Negative deviations occur, as might be expected, when there are large attrac-

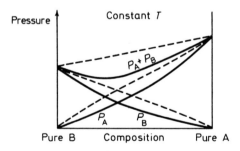

Figure 12.15. Negative deviation from Raoult's law

tive forces between the components. Hence, we would expect the volume of a solution exhibiting negative deviations to be less than the sum of the volumes of the pure components. Examples are acetone–chloroform, pyridine–formic acid solutions in which intermolecular hydrogen-bonding occurs.

Solutions for which the vapour pressures of the components are greater than those expected for the ideal solution are said to exhibit *positive deviations*, as in Figure 12.16. This type of deviation is exhibited if at least one of the components is self-associated. The formation of the solutions is accompanied by expansion on mixing

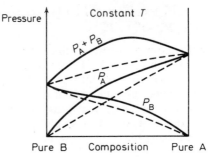

Figure 12.16. Positive deviation from Raoult's law

the pure components, examples being diethyl ether–acetone, heptane–ethanol, water–ethanol.

When deviations from Raoult's law are large, the total vapour pressure of the solution may go through a minimum or a maximum (Figures 12.15 and 12.16).

12.14   The Duhem–Margules equation

Consider a two-component liquid solution in equilibrium with its vapour. The chemical potential of a component must be the same in both phases:

$$\mu_i^{\text{liq}} = \mu_i^{\text{vap}} \tag{12.21}$$

Let us assume that the vapour is a mixture of ideal gases. Since the presence of one ideal gas does not influence the properties of another, the chemical potential of a component is given by Eq. (10.37):

$$\mu_i^{\text{vap}} = \mu_i^{\ominus} + RT \ln \frac{P_i}{P_i^{\ominus}} \tag{12.22}$$

where P_i is the partial pressure of the gas. From Eq. (12.21),

$$\mu_i^{\text{liq}} = \mu_i^{\ominus} + RT \ln \frac{P_i}{P_i^{\ominus}} \tag{12.23}$$

Consider an infinitesimal change in composition at constant temperature and pressure. The change in the chemical potential of i in the liquid phase is given, on differentiating Eq. (12.23) and omitting the superscript, by

$$d\mu_i = RT \, d \ln P_i \tag{12.24}$$

Now, for a change at constant temperature and pressure, the Gibbs–Duhem equation reduces to

$$\Sigma_i x_i \, d\mu_i = 0$$

and for a binary solution of substances A and B, to

$$x_A \, d\mu_A + x_B \, d\mu_B = 0$$

Dividing by dx_A at constant temperature and pressure,

$$x_A \left(\frac{\partial \mu_A}{\partial x_A} \right)_{T,P} + x_B \left(\frac{\partial \mu_B}{\partial x_A} \right)_{T,P} = 0 \tag{12.25}$$

Remembering that for a binary system $x_A = 1 - x_B$ and therefore $dx_A = -dx_B$, and employing Eq. (12.24), we have

$$x_A \left(\frac{\partial \ln P_A}{\partial x_A} \right)_{T,P} = x_B \left(\frac{\partial \ln P_B}{\partial x_B} \right)_{T,P} \tag{12.26}$$

or alternatively,

$$\frac{x_A}{P_A} \left(\frac{\partial P_A}{\partial x_A} \right)_{T,P} = \frac{x_B}{P_B} \left(\frac{\partial P_B}{\partial x_B} \right)_{T,P} \tag{12.27}$$

Both these equations are known as the Duhem–Margules equation. Equation 12.27 relates the slopes of the curves on a diagram of vapour pressure and composition, i.e. it gives the dependence of the gradient of the vapour–pressure curve of A on that of B. These equations are very useful for checking the consistency of experimental data.

It is seen from Figure 12.17 that Eq. (12.27) may be written as

$$\frac{\tan \phi_A}{\tan \theta_A} = \frac{\tan \phi_B}{\tan \theta_B}$$

for all compositions.

Note that the only assumption made in the derivation of the Duhem–Margules equation was that the vapour is a mixture of ideal gases. Deviations from the Duhem–Margules equation due to departures from this assumption are small at relatively low pressures. It is only at high pressures, and when there is association of the molecules in the vapour phase (e.g. acetic acid solutions), that deviations are large.

The Duhem–Margules equation would be precisely obeyed by

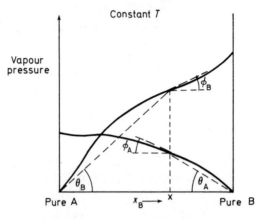

Figure 12.17.

real gases if, instead of using partial pressures in Eq. (12.22), partial fugacities had been used.

12.15   Henry's law

Consider a very dilute solution of A in B such that each A molecule is completely surrounded by B molecules. All the A molecules have a similar environment and, hence, their tendency to escape from the solution is proportional to the mole fraction of A in the solution. If the vapour behaves as an ideal gas, then the vapour pressure is a measure of the escaping tendency. Hence,

$$P_A = k_A x_A \qquad (12.28)$$

where k_A is a constant at a given temperature.

Henry's law states that *the partial vapour pressure of a solute in a dilute solution is proportional to its mole fraction.*

The (solvent) B molecules in the very dilute solution are completely surrounded by other B molecules and, hence, by a similar argument to that given above, we would expect Raoult's law to hold for the solvent.

Figure 12.18 gives the vapour pressures of substances A and B which together form solutions having positive deviations from Raoult's law. The dashed lines show the vapour pressures as predicted by Raoult's law; the dotted lines show the vapour pressures as predicted by Henry's law. As a solution becomes more dilute, the solute obeys Henry's law and the solvent obeys Raoult's law. That is, as $x_A \to 0$ and $x_B \to 1$, $P_A \to k_A x_A$ and $P_B \to x_B P_B^\bullet$. Similarly, as $x_A \to 1$ and $x_B \to 0$, $P_A \to x_A P_A^\bullet$ and $P_B \to k_B x_B$. Solutions exhibiting negative deviations may be discussed similarly.

We have previously stated that for an ideal solution the vapour

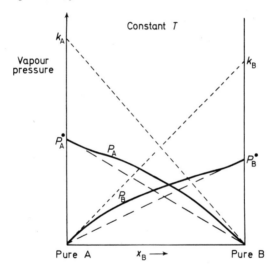

Figure 12.18. Illustration of Raoult's law and Henry's law

pressure of each component is proportional to the mole fraction of that component in solution. Hence, *very dilute solutions are ideal solutions.*

12.16 Raoult's law: composition–pressure and composition–temperature diagrams for two completely miscible liquids

For two substances forming perfect solutions, i.e. solutions obeying Raoult's law over all compositions, the total pressure–composition diagram at constant temperature and temperature–composition diagram at constant pressure are as shown in Figures 12.19 and 12.20, respectively.

According to the phase rule, for a two-component solution ($C = 2$) we have $F + \mathscr{P} = 4$. If one variable is kept fixed (e.g. constant

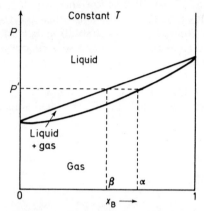

Figure 12.19. *Pressure–composition phase diagram for a binary system obeying Raoult's law*

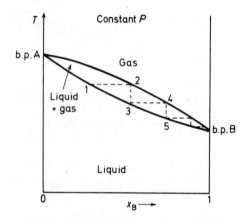

Figure 12.20. *Temperature–composition phase diagram for a binary system obeying Raoult's law*

pressure, Figure 12.20), $F + \mathscr{P} = 3$. For a system consisting of one phase, $\mathscr{P} = 1$ so that $F = 2$, i.e. two functions of state are required to describe the intensive state of the system. For example, in the one-phase regions, both temperature and pressure are required to describe the state of a system (represented by a point). For a two-phase system, $\mathscr{P} = 2$, so that $F = 1$, i.e. only one variable is required to describe the state of the system (see the area bounded by the two curves in Figure 12.19). For example, if the pressure is P', then the composition of the vapour is α and that of the solution is β. Or, if the composition of the vapour is α, then the pressure must be P', and the composition of the solution β, for a two-phase system.

If a liquid is in state 1 (Figure 12.20), then the vapour in equilibrium with it is in state 2. If this vapour is cooled to give liquid in state 3, this liquid will have vapour in state 4 in equilibrium with it, etc. This process is known as fractional distillation. It can be seen that the distillate, i.e. the cooled vapour, becomes richer in component B with each distillation, which must mean that the residue becomes richer in component A. Fractional distillation may therefore be employed to separate the components of a perfect solution.

12.17 Phase diagrams for completely miscible liquids which do not obey Raoult's law

Diagrams analogous to Figures 12.19 and 12.20 are shown in Figures 12.21 and 12.22, for substances showing large positive deviations from Raoult's law, and in Figures 12.23 and 12.24, for

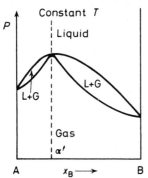

Figure 12.21. Pressure–composition diagram for a binary system showing positive deviation from Raoult's law

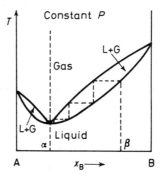

Figure 12.22. Temperature–composition diagram for a binary system showing positive deviation from Raoult's law

substances showing large negative deviations. Solutions of composition other than α, for positive deviations, or other than γ, for negative deviations, boil to give vapours of different compositions. An example is the solution of composition β in Figure 12.22 The mixture of composition α is of minimum boiling point and, similarly, that of composition γ is of maximum boiling point. These mixtures are known as *azeotropic*, or constant-boiling, mixtures.

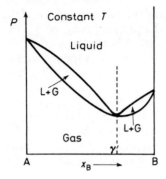

Figure 12.23. Pressure–composition diagram for a binary system showing negative deviation from Raoult's law

Figure 12.24. Temperature–composition diagram for a binary system showing negative deviation from Raoult's law

It can be seen from Figure 12.22 that fractional distillation (going down the steps) for substances showing large positive deviations results in a distillate of composition α. The residue would eventually be pure A. Similarly, from Figure 12.24, fractional distillation results in a distillate of pure A or B, depending on to which side of γ lay the original composition of the solution, the residue being the azeotropic mixture. Hence, substances which form azeotropic mixtures cannot be separated by distillation alone, although, if there is a change in the total pressure, further separation may be obtained.

Consider a system consisting of a two-component solution, in equilibrium with its vapour, undergoing an infinitesimal change. According to the Gibbs–Duhem equation we may write, for the liquid phase,

$$S \, dT - V \, dP + n_A \, d\mu_A + n_B \, d\mu_B = 0$$

and for the gaseous phase,

$$S' \, dT - V' \, dP + n'_A \, d\mu'_A + n'_B \, d\mu'_B = 0$$

where the prime indicates the gaseous phase, and S and V are the total entropy and volume of a given phase. Dividing each equation by the total number of moles in the particular phase,

$$s \, dT - v \, dP + x_A \, d\mu_A + x_B \, d\mu_B = 0$$
$$s' \, dT - v' \, dP + x'_A \, d\mu'_A + x'_B \, d\mu'_B = 0$$

where s and v are the *mean* molar entropy and volume of a given phase. Subtracting these two equations,

$$(s' - s)\,\mathrm{d}T - (v' - v)\,\mathrm{d}P + x'_A\,\mathrm{d}\mu'_A - x_A\,\mathrm{d}\mu_A + x'_B\,\mathrm{d}\mu'_B - x_B\,\mathrm{d}\mu_B = 0$$

Remembering that $x_B = 1 - x_A$ and $x'_B = 1 - x'_A$ and that the solution and vapour are in equilibrium before and after the change, giving $\mathrm{d}\mu_A = \mathrm{d}\mu'_A$ and $\mathrm{d}\mu_B = \mathrm{d}\mu'_B$, we have

$$(s' - s)\,\mathrm{d}T - (v' - v)\,\mathrm{d}P + (x_A - x'_A)(\mathrm{d}\mu_B - \mathrm{d}\mu_A) = 0 \qquad (12.29)$$

Therefore, the dependence of pressure on composition at constant temperature is given by

$$(v' - v)\left(\frac{\partial P}{\partial x_A}\right)_T = (x_A - x'_A)\left(\frac{\partial \mu_B}{\partial x_A} - \frac{\partial \mu_A}{\partial x_A}\right)_T \qquad (12.30)$$

Now for an azeotropic mixture, where the mole fraction of A is the same in the liquid and gaseous phases, i.e. $x_A = x'_A$, the right-hand side of Eq. (12.30) is zero. Since the molar volume of the solution is obviously not equal to the molar volume of its vapour, $v' \neq v$ and, hence,

$$\left(\frac{\partial P}{\partial x_A}\right)_T = 0$$

This is, of course, the mathematical condition for a maximum or minimum vapour pressure which is observed in Figures 12.21 and 12.23 for the azeotropic mixtures.

From Eq. (12.29) the temperature dependence on composition at constant pressure is given by

$$(s' - s)\left(\frac{\partial T}{\partial x_A}\right)_P = (x'_A - x_A)\left(\frac{\partial \mu_B}{\partial x_A} - \frac{\partial \mu_A}{\partial x_A}\right)_P$$

As before, for an azeotropic mixture $x_A = x'_A$, so the right-hand side is zero. Since, in general, $s \neq s'$ we have

$$\left(\frac{\partial T}{\partial x_A}\right)_P = 0$$

Therefore the boiling temperature of an azeotropic mixture is a maximum or a minimum at a given pressure (see Figures 12.24 and 12.22).

Although the above discussion has dealt with liquid–vapour

systems, it is equally applicable to solid–liquid systems which also, of course, form azeotropes.

12.18   Pressure–temperature diagrams

It is sometimes useful to plot a pressure–temperature diagram at constant total composition for a binary two-phase system (Figure 12.25).

The temperature at which a binary liquid mixture of given composition boils at different pressures is indicated by the *vaporisation curve*, which is also known as the *bubble-point curve*. The

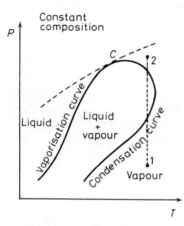

Figure 12.25. P–T diagram for a binary system

temperature at which a gaseous mixture condenses at different pressures is indicated by the *condensation curve*, which is also known as the *dew-point curve*. At the point C, known as the *plait point*, the two phases have the same composition. The plait point resembles the critical point of a pure substance, and is also sometimes known as the critical point of the solution. Note that it need not correspond to the highest temperature, or highest pressure, at which the liquid and gaseous phases are in equilibrium.

The dotted line through C shows how the plait point varies with overall composition (see also Figure 12.26). Above this line, there is no distinction between the liquid and gaseous phases.

In Figure 12.26, C_A and C_B are the critical points of the pure substances A and B of the mixture. The maximum temperature, maximum pressure and plait points for mixtures of various compositions are illustrated.

It is interesting to consider the result of an isothermal compression of a gaseous mixture from state 1 to state 2 (Figure 12.25). It is seen that, with increase in pressure, the system passes from states involving the vapour phase to states in which vapour and liquid are in equilibrium, and finally back to states involving the gaseous phase.

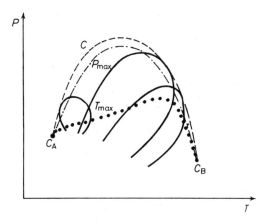

Figure 12.26. Variation of the plait point, the maximum pressure and maximum temperature with composition

*12.19 Two binary phases in equilibrium

At equilibrium, the chemical potential of component A is the same in the two phases;

$$\mu_A = \mu'_A \qquad (12.31)$$

where the prime is used to distinguish between the phases. Differentiating,

$$d\mu_A = d\mu'_A$$

this being true for an infinitesimal change at the end of which the two phases are still in equilibrium. Employing Eq. (10.38), we have

$$\bar{V}_A\,dP - \bar{S}_A\,dT + \left(\frac{\partial\mu_A}{\partial x_A}\right)_{P.T} dx_A = \bar{V}'_A\,dP - \bar{S}'_A\,dT + \left(\frac{\partial\mu'_A}{\partial x'_A}\right)_{P.T} dx'_A$$

Or,

$$(\bar{V}_A - \bar{V}'_A)\,dP - (\bar{S}_A - \bar{S}'_A)\,dT = -\left(\frac{\partial\mu_A}{\partial x_A}\right)_{P.T} dx_A + \left(\frac{\partial\mu'_A}{\partial x'_A}\right)_{P.T} dx'_A$$

$$(12.32)$$

Similarly, for component B,

$$(\bar{V}_B - \bar{V}_B')\,dP - (\bar{S}_B - \bar{S}_B')\,dT = -\left(\frac{\partial\mu_B}{\partial x_B}\right)_{P,\,T}dx_B + \left(\frac{\partial\mu_B'}{\partial x_B'}\right)_{P,\,T}dx_B' \tag{12.33}$$

Multiplying Eq. (12.32) by x_A' and Eq. (12.33) by x_B', adding and using Eq. (12.25) for both phases, remembering that $dx_A' = -dx_B'$ and $dx_A = -dx_B$, we obtain

$$[x_A'(\bar{V}_A - \bar{V}_A') + x_B'(\bar{V}_B - \bar{V}_B')]\,dP - [x_A'(\bar{S}_A - \bar{S}_A') + x_B'(\bar{S}_B - \bar{S}_B')]\,dT$$

$$= \frac{(x_A - x_A')}{x_B}\left(\frac{\partial\mu_A}{\partial x_A}\right)_{P,\,T}dx_A \tag{12.34}$$

It is seen from this equation that of the three variables P, T and x_A only two are independent. This is in agreement with the phase rule, $F + \mathscr{P} = C + 2$, for, since $C = 2$ and $\mathscr{P} = 2$, we have $F = 2$.

The same results as were derived from Eq. (12.29) regarding azeotropic mixtures may be derived similarly from Eq. (12.34), i.e. if either $(\partial P/\partial x_A)_T$ and $(\partial T/\partial x_A)_P$ are zero, then $x_A = x_A'$.

If x_A is maintained constant, i.e. the composition of that phase is kept constant, then

$$\left(\frac{\partial P}{\partial T}\right)_{x_A} = \frac{x_A'(\bar{S}_A - \bar{S}_A') + x_B'(\bar{S}_B - \bar{S}_B')}{x_A'(\bar{V}_A - \bar{V}_A') + x_B'(\bar{V}_B - \bar{V}_B')} \tag{12.35}$$

Since Eq. (12.31) holds, Eq. (10.35) gives

$$\bar{H}_A - T\bar{S}_A = \bar{H}_A' - T\bar{S}_A' \tag{12.36}$$

and similarly for component B. Eq. (12.35) may be written as

$$\left(\frac{\partial P}{\partial T}\right)_{x_A} = \frac{x_A'(\bar{H}_A - \bar{H}_A') + x_B'(\bar{H}_B - \bar{H}_B')}{T[x_A'(\bar{V}_A - \bar{V}_A') + x_B'(\bar{V}_B - \bar{V}_B')]} = \frac{\Delta H}{T\Delta V} \tag{12.37}$$

It can be seen that ΔH is the heat exchanged when x_A' moles of A and x_B' moles of B are transferred from the primed phase to the unprimed phase, the composition of the latter phase not being altered. We can imagine this phase as being so large that the addition of x_A' and x_B' does not alter its composition. ΔH is known as the differential heat of solution. Similarly, ΔV is the volume change accompanying the exchange of x_A' and x_B' moles. It is interesting to compare Eq. (12.37) with the Clapeyron equation (Eq. 12.5), which applies to a pure substance.

For liquid–gaseous (or solid–gaseous) phase equilibria, the molar volume of the liquid (or solid) is negligible compared to that of the gas. If we assume the gas phase to be a mixture of ideal gases, then, using the ideal-gas equation of state,

$$\left(\frac{\partial \ln P}{\partial T}\right)_{x_A} = \frac{\Delta H}{RT^2}$$

12.20 Eutectic point

The temperature–composition diagram at constant pressure for a binary system of A and B, which are completely miscible in the liquid phase but immiscible in the solid phase, is given in Figure 12.27. The temperatures L and Q are the freezing points of pure A and B at this particular pressure. The point E is known as the eutectic point. The eutectic mixture has the lowest freezing point.

Consider the isobaric cooling of the liquid solution of composition K. At the point M, pure solid A starts to separate from solution.

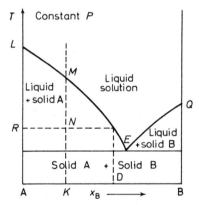

Figure 12.27. System having a eutectic point

Further reduction in temperature results in more A solidifying and the solution therefore becoming richer in B. At the point N, solid A and liquid of composition D are in equilibrium. At the eutectic temperature, solids A and B are in equilibrium with liquid of the eutectic composition. Further cooling results in the eutectic mixture freezing. At one time, it was thought that the eutectic mixture was a compound. However, under a microscope distinct crystals of A and B are observed. Also, if the pressure is altered, the composition of the eutectic mixture is different.

It can be seen from the curve *LE* that addition of B to pure A results in a lower freezing point. Hence *LE* may be regarded as a *freezing-point curve for* A. It may also be regarded as the *solubility curve of* A *in* B. *LE* shows how the composition of the solution saturated in A varies with temperature. Basically, for a freezing-point or solubility curve we have a system in which a pure solid is in equilibrium with its solution. *EQ* may be similarly regarded as a freezing-point curve or a solubility curve for B.

Now according to the phase rule for equilibrium in a two-component system ($C = 2$), $F + \mathscr{P} = 4$ and, for constant pressure, $F + \mathscr{P} = 3$. For a one-phase system $\mathscr{P} = 1$, so that $F = 2$ – that is, two variables are required to specify the state of the solution (i.e. two coordinates are required to locate a point in the one-phase region). For $\mathscr{P} = 2$, $F = 1$ – that is, only one intensive variable is required to specify the intensive state of the system. For example, for a system rich in A, if the temperature is R, then it is known that solid A is in equilibrium with solution of composition D, or, if the composition of the solution is D, then the temperature must be R.

At the eutectic temperature, three phases coexist, so that $\mathscr{P} = 3$ and hence $F = 0$. That is, there are no degrees of freedom, meaning that the state of the system is completely specified at *constant pressure* on stating that solid A and solid B are in equilibrium with the liquid phase.

*12.21   Thermodynamic treatment of the eutectic point

Solids A and B are in equilibrium with their solution. The chemical potential of A in solution is equal to that of A in the solid phase:

$$\mu_A = \mu_A'$$

the prime indicating the solid phase.

Differentiating, i.e. considering an infinitesimal change that results in equilibrium, we have

$$d\mu_A = d\mu_A'$$

μ_A may be expressed in terms of P, T and x_A. μ_A' may be expressed in terms of P and T only. Hence,

$$\left(\frac{\partial \mu_A}{\partial P}\right)_{T, x_A} dP + \left(\frac{\partial \mu_A}{\partial T}\right)_{P, x_A} dT + \left(\frac{\partial \mu_A}{\partial x_A}\right)_{P, T} dx_A$$

$$= \left(\frac{\partial \mu_A'}{\partial P}\right)_T dP + \left(\frac{\partial \mu_A'}{\partial T}\right)_P dT$$

or,

$$\bar{V}_A \, dP - \bar{S}_A \, dT + \left(\frac{\partial \mu_A}{\partial x_A}\right)_{P, T} dx_A = \bar{V}'_A \, dP - \bar{S}'_A \, dT$$

Rearranging,

$$(\bar{V}'_A - \bar{V}_A) \, dP - (\bar{S}'_A - \bar{S}_A) \, dT = \left(\frac{\partial \mu_A}{\partial x_A}\right)_{P, T} dx_A \qquad (12.38)$$

Similarly, for component B,

$$(\bar{V}'_B - \bar{V}_B) \, dP - (\bar{S}'_B - \bar{S}_B) \, dT = \left(\frac{\partial \mu_B}{\partial x_B}\right)_{P, T} dx_B \qquad (12.39)$$

On multiplying Eq. (12.38) by x_A and Eq. (12.39) by x_B and adding, the right-hand side is zero according to Eq. (12.25), so that we obtain

$$\frac{dP}{dT} = \frac{x_A(\bar{S}'_A - \bar{S}_A) + x_B(\bar{S}'_B - \bar{S}_B)}{x_A(\bar{V}'_A - \bar{V}_A) + x_B(\bar{V}'_B - \bar{V}_B)}$$

Employing Eq. (12.36),

$$\frac{dP}{dT} = \frac{x_A(\bar{H}'_A - \bar{H}_A) + x_B(\bar{H}'_B - \bar{H}_B)}{T[x_A(\bar{V}'_A - \bar{V}_A) + x_B(\bar{V}'_B - \bar{V}_B)]} = \frac{\Delta H}{T \Delta V}$$

It is seen from this equation that the only variables are P and T. Hence, at a given pressure, the state of the eutectic mixture is determined.

ΔH is the heat exchanged when x_A moles of pure A and x_B moles of pure B are formed from one mole of solution, and ΔV is the accompanying volume change.

Problems

1. The latent heat of vaporisation of ether is $27 \cdot 33 \, \text{kJ mol}^{-1}$ and its boiling point is $307 \cdot 7$ K. Calculate (a) dP/dT near its boiling point, (b) the vapour pressure at $309 \cdot 2$ K, and (c) the temperature at which the vapour pressure is $100\,000 \, \text{N m}^{-2}$.
Answer (a) $35\,400 \, \text{N m}^{-2} \, \text{K}^{-1}$, (b) $106\,600 \, \text{N m}^{-2}$, (c) $307 \cdot 3$ K
2. Given that $dP/dT = 361 \cdot 4 \, \text{N m}^{-2} \, \text{K}^{-1}$, and assuming that steam behaves as an ideal gas, calculate the latent heat of vaporisation ($R = 8 \cdot 314 \, \text{J K}^{-1} \, \text{mol}^{-1}$).
Answer $40 \cdot 5 \, \text{kJ mol}^{-1}$

3. Calculate the latent heat of fusion if $dP/dT = 13.5 \times 10^6$ N m^{-2} K^{-1} and the molar volumes of ice and water are 19.63×10^{-6} and 18.00×10^{-6} m^3 mol^{-1}.

Answer   601.7 J mol^{-1}

4. Find the latent heat of vaporisation of benzene graphically from the following vapour pressure data:

T/K	273	283	293	303	313
$P/N\,m^{-2}$	3 530	6 050	9 940	15 310	25 200

Answer 29.25 J mol^{-1}

5. If the vapour pressures of ice at 252.7 and 253.7 K are 97.9 and 107.7 N m^{-2}, respectively, calculate the latent heat of sublimation.

Answer   50.4 kJ mol^{-1}

6. At 298 K, the vapour pressure of mercury is 0.44 N m^{-2}. Assuming that Trouton's rule is obeyed, calculate the boiling point of mercury.

Answer   630 K

7. Derive the phase rule from first principles.

8. The measured vapour pressures of ethanol and water in a two-component solution at 298 K at various mole fractions of ethanol are given below:

x(ethanol)	P(ethanol)/N m^{-2}	P(water)/N m^{-2}
0.3	4160	2585
0.4	4560	2445
0.5	4920	2305
0.6	5340	2103

Show the validity of the Duhem–Margules equation at x(ethanol) = 0.4 and 0.5.

9. The vapour pressure of a liquid A is given by

$$\log_{10}P = 10.69 - \frac{2120}{T}$$

and that of the solid A by

$$\log_{10}P = 2.115 - \frac{31.43}{T}$$

where P is in N m^{-2}.

Calculate (a) the triple point, (b) the latent heat of fusion, and (c) the latent heat of vaporisation.

Answer (a) 243.5 K, (b) 602 J mol^{-1}, (c) 40.61 kJ mol^{-1}

10. Since for a lambda transition $\Delta S = 0$ and $\Delta V = 0$, dP/dT is

indeterminate, show that in fact

(a) $\dfrac{\mathrm{d}P}{\mathrm{d}T} = -\dfrac{\left(\dfrac{\partial \Delta V}{\partial T}\right)_P}{\left(\dfrac{\partial \Delta V}{\partial P}\right)_T}$, (b) $\dfrac{\mathrm{d}P}{\mathrm{d}T} = \dfrac{\Delta C_P}{T\left(\dfrac{\partial \Delta V}{\partial T}\right)_P}$

and, hence, (c) $\Delta C_P = -\dfrac{T\left(\dfrac{\partial \Delta V}{\partial T}\right)_P^2}{\left(\dfrac{\partial \Delta V}{\partial P}\right)_T}$

13

Solutions

13.1 Chemical potential and the perfect gas mixture

Consider a mixture of ideal gases (i.e. a perfect gas solution). For a
pure ideal gas we have shown (Eq. 11.3) that

$$\mu = \mu^{\ominus} + RT \ln \frac{P}{P^{\ominus}} \tag{13.1}$$

Now, according to the Gibbs–Dalton law (Section 6.1), an ideal gas
in a gaseous mixture behaves as if it alone occupied the entire
volume of the system at the temperature of the system. That is, each
component of a mixture of ideal gases behaves independently of
the other components. Therefore, from Eq. (13.1), the chemical
potential of component i in the mixture is

$$\mu_i = \mu_i^{\ominus} + RT \ln \frac{P_i}{P_i^{\ominus}} \tag{13.2}$$

where P_i and $P_i^{\ominus}$ are the partial pressures of the gas in the mixture
and in the standard state.

The standard state I of a gas is usually defined as *the state of the
pure gas at a fugacity of* 101 325 N m^{-2} *and at the temperature under
consideration, the gas behaving ideally* (see Section 11.4). (For an
ideal gas the property fugacity may be substituted by pressure.)

A gas mixture that obeys the Gibbs–Dalton law over all compositions is known as a *perfect gas mixture*. From the Gibbs–Dalton
law, it is obvious that for such mixtures $\mu_i = g_i$ and $\mu_i^{\ominus} = g_i^{\ominus}$.
Making use of Eq. (6.8), which states

$$P_i = x_i P$$

where P is the total pressure and x_i is the mole fraction of component i, Eq. (13.2) becomes

$$\mu_i = \mu_i^\ominus + RT \ln \frac{P}{P_i^\ominus} + RT \ln x_i \qquad (13.3)$$

From Eq. (13.1), it is seen that $\mu_i^\ominus + RT \ln P/P_i^\ominus$ is the chemical potential of i at the temperature T and pressure P, which we shall denote by $\mu_i^\bullet$. Hence,

$$\mu_i = \mu_i^\bullet + RT \ln x_i \qquad (13.4)$$

If we were more interested in the measurement of mole fraction than of partial pressure, then we would use this equation rather than Eq. (13.2). If we do make use of Eq. (13.4), then it is more convenient to define another state as the standard state in order to simplify our working. *The standard state II of a gas in a mixture is defined as the state of the pure gas at the temperature and pressure of the mixture under consideration.* The standard chemical potential, $\mu_i^\bullet$, as we have already said, is the chemical potential of pure gas i at the temperature T and pressure P. In other words, the value of $\mu^\bullet$ depends on both temperature and pressure, whereas the value of the standard chemical potential $\mu_i^\ominus$ in Eq. (13.2) depends only on temperature, since it is always referred to a pressure of 101 325 N m^{-2}. We shall employ the superscript $^\bullet$ to indicate standard states which depend on both temperature and pressure.

13.2 Mixtures of non-ideal gases

Let us consider a mixture of real gases. On comparing Eq. (13.2) with Eq. (13.1), and bearing in mind the concept of fugacity (Eq. 11.9), we may define the fugacity of a component i in a real gas mixture for an *isothermal change* by

$$d\mu_i = RT \, d \ln f_i \qquad (13.5)$$

On integration of this equation, an arbitrary integration constant would appear. Therefore this equation is only sufficient to define fugacity if we specify further that, as the pressure of a gas mixture is decreased, the fugacity of component i tends to the partial pressure of i, i.e. as $P \to 0$, then $f_i \to P_i$. In other words, the mixture tends to ideal behaviour as the pressure is decreased.

Integrating Eq. (13.5) from the standard state to the state of the mixture,

$$\int_{\mu_i^{\ominus}}^{\mu_i} d\mu_i = RT \int_{f_i^{\ominus}}^{f_i} d \ln f_i$$

Hence,

$$\mu_i = \mu_i^{\ominus} + RT \ln \frac{f_i}{f_i^{\ominus}} \qquad (13.6)$$

The standard state may, for example, be defined as for a real gas (see page 252) – that is, the state of the pure gas at a fugacity of 101 325 $N\ m^{-2}$ at the temperature under consideration, assuming the gas to behave ideally.

The *activity of component* i, a_i, in the mixture is defined as the ratio of the fugacity of component i in the mixture to the fugacity of component i in its standard state.

$$a_i = \frac{f_i}{f_i^{\ominus}} \qquad (13.7)$$

Hence, the value of activity is dependent on the definition of the standard state. From Eq. (13.6),

$$\mu_i = \mu_i^{\ominus} + RT \ln a_i \qquad (13.8)$$

In the general case when the standard state depends on *pressure* as well as temperature, we have $a_i = f_i/f_i^{\bullet}$ and therefore

$$\mu_i = \mu_i^{\bullet} + RT \ln a_i \qquad (13.9)$$

*13.3   Determination of fugacity

The determination of the fugacity of a component i of an imperfect gas mixture follows similar ideas to those employed for a pure gas (see Section 11.5). Equation (10.32) states:

$$\left(\frac{\partial \mu_i}{\partial P}\right)_{T,n} = \bar{V}_i$$

Employing Eq. (13.5), we have, for an isothermal process at constant composition,

$$RT\ d \ln f_i = \bar{V}_i\ dP$$

Subtracting $RT\,\mathrm{d}\ln P_i$ from both sides of the equation,

$$RT\,\mathrm{d}\ln\frac{f_i}{P_i} = \bar{V}_i\,\mathrm{d}P - RT\,\mathrm{d}\ln P_i$$

Using Dalton's law and substituting $\phi_i = f_i/P_i$,

$$RT\,\mathrm{d}\ln\phi_{i} = \bar{V}_i\,\mathrm{d}P - RT\,\mathrm{d}\ln P - RT\,\mathrm{d}\ln x_i$$

Since the composition is constant,

$$RT\,\mathrm{d}\ln\phi_i = \left(\bar{V}_i - \frac{RT}{P}\right)\mathrm{d}P$$

Integrating from zero pressure to the pressure P, and remembering that as $P \to 0, f_i \to P_i$, i.e. $\phi_i \to 1$,

$$\ln\phi_i = \int_0^P \left(\frac{\bar{V}_i}{RT} - \frac{1}{P}\right)\mathrm{d}P$$

To evaluate the fugacity coefficient ϕ_i of component i at temperature T and pressure P from the above equation, the variation of the partial molar volume $\bar{V}_i$ with pressure at the temperature T must be known. To determine each experimental value of $\bar{V}_i$ over the integration range, one of the methods in Section 10.7 must be used. This is usually a rather lengthy procedure. The fugacity may then be determined using $f_i = \phi_i P_i$.

For a two-component mixture, the values obtained for the fugacities of components A and B may be checked using the Gibbs–Duhem equation at constant temperature and pressure,

$$n_\mathrm{A}\,\mathrm{d}\mu_\mathrm{A} + n_\mathrm{B}\,\mathrm{d}\mu_\mathrm{B} = 0$$

Substituting for $\mathrm{d}\mu$ from Eq. (13.5),

$$\mathrm{d}\ln f_\mathrm{A} = -\frac{n_\mathrm{B}}{n_\mathrm{A}}\cdot\mathrm{d}\ln f_\mathrm{B}\ \ = -\frac{x_\mathrm{B}}{x_\mathrm{A}}\,\mathrm{d}\ln f_\mathrm{B}$$

or,

$$\mathrm{d}\ln f_\mathrm{A} = -\frac{x_\mathrm{B}}{1 - x_\mathrm{B}}\,\mathrm{d}\ln f_\mathrm{B}$$

Hence, from a knowledge of the variation of f_B with concentration, f_A may be calculated by graphical integration of the graph of $x_\mathrm{B}/1 - x_\mathrm{B}$ against $\ln f_\mathrm{B}$.

13.4   Solutions

The concepts used for gaseous solutions may be extended to all solutions – gas, liquid, or solid; ideal or non-ideal – so that for *any* solution the chemical potential of each component at a given temperature is given by

$$\mu_i = \mu_i^{\ominus} + RT \ln \frac{f_i}{f_i^{\ominus}} \tag{13.10}$$

or by

$$\mu_i = \mu_i^{\bullet} + RT \ln \frac{f_i}{f_i^{\bullet}} \tag{13.11}$$

where $^{\ominus}$ indicates a standard state independent of pressure, and $^{\bullet}$ indicates a standard state dependent on pressure.

13.5 Definition of an ideal solution

We shall define an ideal solution (gas, liquid or solid) as one for which the fugacity of each component is proportional to the mole fraction of that component in the solution, i.e. the fugacity of *each* component is given by

$$\boxed{f_i = x_i f_i^{\bullet}} \tag{13.12}$$

where x_i is the mole fraction of component i in the solution and $f_i^{\bullet}$ is *a constant* for a given temperature and pressure.

If the solution remains ideal over all possible compositions, even approaching $x_i = 1$, then $f_i^{\bullet}$ is the fugacity of the pure component i at the same temperature and pressure as the solution. A solution which remains *ideal over all compositions* is, of course, a *perfect* solution. The temperature, pressure and composition range over which a particular real solution behaves ideally is a matter for experimental determination.

AN IDEAL GAS SOLUTION

An imperfect gas mixture may be an ideal gas solution, since for such mixtures Eq. (13.12) may hold even to quite high pressures. In other words, an ideal gas solution need not be a mixture of ideal gases. For an ideal gas solution, the fugacity of each component may be computed from a knowledge of the fugacity of the pure component at the same temperature and pressure, using Eq. (13.12).

If we consider that the components of the solution behave as ideal gases, then Eq. (13.12) becomes

$$P_i = x_i P$$

which is merely a statement of Dalton's law. Hence, a perfect gas mixture is a special case of an ideal gas solution.

For a liquid solution, in Eq. (13.12) $f_i^\bullet$ is a constant at a given T and P.

(1) Consider a liquid solution in equilibrium with its vapour, the latter behaving as a mixture of ideal gases. Then, from Section 11.8, $f_i^{\text{liq}} = P_i^{\text{vap}}$. Hence,

$$P_i^{\text{vap}} = x_i f_i^\bullet$$

This is merely a statement of *Henry's law* — see Eq. (12.28), where $f_i^\bullet \equiv k_i$. Therefore Eq. (13.12) is a more general statement of Henry's law. In other words, a liquid solution whose components obey Henry's law is an ideal solution.

If we plot the experimental vapour pressure or, more strictly, the fugacity of a component i in a liquid (or solid) solution against its mole fraction (Figure 13.1), we find that

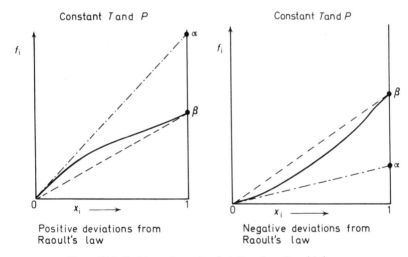

Constant T and P Constant T and P

Positive deviations from Raoult's law   Negative deviations from Raoult's law

Figure 13.1. Positive and negative deviations from Raoult's law

the component i obeys Henry's law only in dilute solution. If Henry's law were obeyed over the entire range of concentration, the fugacity of i would lie on the dashed and dotted line. The point α gives the value of $f_i^\bullet$ according to Henry's law, i.e. it gives the fugacity of component i in the hypothetical pure state which would be expected if Henry's law held over the entire range of composition at the temperature and pressure of the solution.

(2) Consider a liquid solution in equilibrium with its vapour, the latter behaving as an ideal gas mixture, so that $f_i^{liq} = P_i^{vap}$. Now consider that the solution remains ideal over all compositions. Then the fugacity of liquid i in the pure state at the temperature and pressure of the solution equals the vapour pressure of the pure component i at the same temperature and pressure: $f_i^{\bullet liq} = P_i^{\bullet vap}$. Hence,

$$P_i^{vap} = x_i P_i^{\bullet vap}$$

which is merely a statement of Raoult's law. That is, Eq. (13.12) is a more general statement of Raoult's law. Therefore a solution obeying Raoult's law is an ideal solution.

From Figure 13.1 it is seen that at high concentrations of component i Raoult's law is obeyed. The dashed line shows the fugacity of component i predicted by Raoult's law and the point β therefore gives the value of $f_i^{\bullet}$ according to this law.

For any solution (gas, liquid or solid), it is seen from Eq. (13.12) that if $x_i = 1$ then $f_i = f_i^{\bullet}$. If the solution tends to ideal behaviour as $x_i \to 1$, then $f_i^{\bullet}$ is the fugacity of pure component i at the temperature and pressure of the solution (see point β). If the solution does *not* tend to ideal behaviour as $x_i \to 1$, then $f_i^{\bullet}$ may be regarded as the fugacity of i at the temperature and pressure of the solution in a hypothetical pure state, this being the pure state which would be achieved if the solution remained ideal as $x_i \to 1$. For example, if we consider that a component i of a solution follows ideal behaviour when obeying Henry's law, then $f_i^{\bullet}$ is given by the point α.

13.6   Ideal solutions—Chemical potential

Differentiating Eq. (13.12) at a given temperature and pressure, we have

$$d \ln f_i = d \ln x_i$$

Using the differential form of Eq. (13.10), or Eq. (13.11), at constant T and P.

$$d\mu_i = RT \, d \ln x_i$$

Hence, integrating from the state where $x_i = 1$ which we shall call the standard state (and shall define explicitly later), which may or may not be pressure-dependent, to the state where the mole fraction is x_i, we have

$$\mu_i = \mu_i^{\bullet} + RT \ln x_i \qquad (13.13)$$

Alternatively, substituting for f_i from Eq. (13.12), Eq. (13.10) gives, for each component of an ideal solution at a given temperature and pressure,

$$\mu_i = \mu_i^{\ominus} + RT \ln \frac{x_i f_i^{\bullet}}{f_i^{\ominus}} \qquad (13.14)$$

where $f_i^{\ominus}$ is the fugacity of the pure component i in the standard state (independent of pressure) and $f_i^{\bullet}$ is the fugacity of i in the pure (and possibly hypothetical) state at the temperature and pressure of the solution. Rearranging Eq. (13.14),

$$\mu_i = \mu_i^{\ominus} + RT \ln \frac{f_i^{\bullet}}{f_i^{\ominus}} + RT \ln x_i$$

Hence, for an ideal solution at a given temperature and pressure, we may write

$$\mu_i = \mu_i^{\bullet} + RT \ln x_i \qquad (13.13)$$

where

$$\mu_i^{\bullet} = \mu_i^{\ominus} + RT \ln \frac{f_i^{\bullet}}{f_i^{\ominus}}$$

the value of $\mu_i^{\bullet}$ depending only on temperature and pressure. As may be seen directly from Eq. (13.13), if $x_i = 1$, $\mu_i^{\bullet}$ is the chemical potential of pure component i at the temperature and pressure of the solution.

In the special case of a mixture of ideal gases, Eq. (13.14) becomes identical to Eq. (13.3).

13.7 Non-ideal solutions

A non-ideal solution is one for which Eq. (13.12) does *not* hold for at least one component. Of course, Eq. (13.10), which states that

$$\mu_i = \mu_i^{\ominus} + RT \ln \frac{f_i}{f_i^{\ominus}}$$

applies to any solution, ideal or otherwise. Once again, we are driven

by the simplicity of the equations in the previous sections to introduce a correction factor γ_i known as the *activity coefficient* of
component i into Eq. (13.12) so that it may also apply to non-ideal
solutions. We therefore write:

$$f_i = \gamma_i x_i f_i^{\bullet} \qquad (13.15)$$

where γ_i is dependent on the temperature, pressure *and composition*
of the solution. Obviously, *for an ideal solution* $\gamma_i = 1$.
    The *activity* of component i is defined by

$$a_i = \gamma_i x_i \qquad (13.16)$$

or by

$$a_i = \frac{f_i}{f_i^{\bullet}} \qquad (13.17)$$

and we see that activity is a dimensionless quantity.
    From Eqs (13.10) and (13.17),

$$\mu_i = \mu_i^{\ominus} + RT \ln \frac{f_i^{\bullet}}{f_i^{\ominus}} + RT \ln a_i$$

or,

$$\mu_i = \mu_i^{\bullet} + RT \ln a_i \qquad (13.18)$$

where $\mu_i^{\bullet}$ is defined as before.
    Comparing Eq. (13.18) with Eq. (13.13), it is obvious that for an
ideal solution $a_i = x_i$, as may also be seen from Eq. (13.16).
    Before proceeding any further, we must discuss standard states
in more detail.

13.8   The standard states of components of a mixture

It would be very elegant, and it would certainly avoid confusion, if
for a given substance, no matter whether a component of a mixture
or not, and no matter what its phase, we selected only one standard
state. However, in practice it is much more convenient to choose

standard states which are best suited to a particular problem. When examining literature, extreme caution must therefore be exercised to note the standard states employed. We give below the commonest standard states in use. Of course, we can always devise our own if these are not suitable.

THE STANDARD STATES OF GASES

The *standard state I* of a gas may be defined strictly as *the state of the pure gas at a fugacity of* 101 325 N m^{-2} *at the temperature under consideration, the gas behaving ideally* (see Figure 11.2). However, if we use Eq. (13.4) or Eq. (13.13), then it is more convenient to define the *standard state II* of a gas as *the state of the pure gas at the temperature and pressure of the mixture.*

THE STANDARD STATES OF LIQUIDS (OR SOLIDS)

The standard states of liquids and solids are similar and will therefore be dealt with together. It is usually convenient for liquid (or solid) solutions if we select different standard states depending on whether the substance is a solvent or solute. For a solvent, the standard state is based on Raoult's law and, for a solute, on Henry's law.

The standard state III of a solvent is *the state of the pure liquid (or solid) at a pressure of* 101 325 N m^{-2} *and at the temperature under consideration.* If, in Figure 13.1, the total pressure is 101 325 N m^{-2}, then the point β illustrates the standard state of component i if it is to be considered as a solvent.

The standard state IV of a solute is *the hypothetical pure state of the liquid (or solid), assuming that it obeys Henry's law over the entire range of composition at a total pressure of* 101 325 N m^{-2} *and at the temperature under consideration.* If, in Figure 13.1, the total pressure is 101 325 N m^{-2}, then the point α illustrates the standard state of component i if it is to be considered as a solute.

Obviously, in the standard state the activity of a component is unity.

It is also quite common to define the standard states of liquids and solids not at a total pressure of 101 325 N m^{-2} but in such a manner that they are pressure-dependent. For example, we might define *the standard state IIIa of a solvent* as *the state of the pure liquid (or solid) under its own vapour pressure at the temperature under consideration,* or the *standard state IIIb of a solvent* as *the state of the pure liquid (or solid) at the temperature and pressure of the solution under consideration.*

Since at relatively low total pressures the vapour pressure of a liquid (or solid) is not affected by pressure, i.e. at low pressures the fugacity in the pure state at a particular pressure and temperature is approximately equal to the fugacity at $101\,325\,\text{N m}^{-2}$ and the same temperature, $f_i^{\bullet} = f_i^{\ominus}$ and these three types of standard state for a solvent will be considered equivalent, and similarly for solutes. That is, at low pressures we shall consider it immaterial when dealing with liquids and solids whether the pressure in the standard state is the vapour pressure, the pressure of the solution or $101\,325\,\text{N m}^{-2}$. In fact, if required, Eq. (11.31) may be used to calculate the change in fugacity between the two standard states.

We have stated above that for liquids and solids at relatively low pressures we may write $f_i^{\bullet} = f_i^{\ominus}$. Therefore Eq. (13.13) reduces approximately to

$$\mu_i = \mu_i^{\ominus} + RT \ln x_i \qquad (13.19)$$

MOLALITY

Unfortunately, it is necessary to consider yet another standard state, which is based on a measure of composition known as *molality*. This is generally used when dealing with aqueous solutions to measure electrolyte compositions.

We have defined the mole fraction of a component A as the ratio of the amount of component A in a solution to the total amount of all the components in the solution,

$$x_A = \frac{n_A}{\Sigma_i n_i} \qquad (13.20)$$

The molality of a solute A is the number of moles of A per kilogramme of solvent.

Consider a solution containing n_A moles of solute A and n_0 moles of solvent of molecular weight M_0. The mass in grammes of the solvent is $n_0 M_0$ and the molality of A, m_A, is given by

$$m_A = \frac{1000 n_A}{n_0 M_0}$$

Multiplying top and bottom by the total number of moles in the solution, $\Sigma_i n_i$, and employing Eq. (13.20),

$$m_A = \frac{1000\, \Sigma_i n_i}{n_0 M_0} \cdot x_A$$

If the solution is *very dilute*, $\Sigma_i n_i$ is approximately n_0 so that

$$m_A \doteqdot \frac{1000}{M_0} \cdot x_A$$

That is, for a very dilute solution, the molality of any solute is approximately proportional to its mole fraction.

Since very dilute solutions tend to ideal behaviour, the solutes in the solution following Henry's law and the solvent following Raoult's law, we may write, for a solute,

$$f_i = m_i f_i^{\bullet m} \tag{13.21}$$

where $f_i^{\bullet m}$ is a constant dependent on temperature and pressure. Hence, for very dilute solutions, we may, for a solute, derive an equation similar to Eq. (13.13) — that is,

$$\mu_i = \mu_i^{\bullet m} + RT \ln m_i \tag{13.22}$$

where $\mu_i^{\bullet m}$ depends on temperature and pressure and is the chemical potential of component i for unit molality.

As before, for non-ideal solutions, i.e. solutions where the solutes

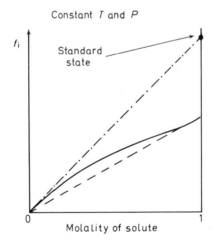

Figure 13.2. Standard states

do not obey Eq. (13.21), we introduce a correcting factor, the activity coefficient γ_{mi}, so that

$$\mu_i = \mu_i^{\bullet m} + RT \ln \gamma_{mi} m_i \tag{13.23}$$

the activity being given by

$$a_{mi} = \gamma_{mi} m_i \tag{13.24}$$

The standard state V for a solute is the hypothetical state of unit molality of the solute, assuming Henry's law to be obeyed, at a pressure of $101\,325\,N\,m^{-2}$ and at the temperature under consideration.

In the graph of fugacity against composition in Figure 13.2, this standard state is illustrated. The activity of a component in its standard state is unity.

13.9　The importance of standard states

The value of the activity of each component of a solution will depend on the standard state chosen for that component, and therefore we must be careful to note which standard state has been employed. This is really another way of deciding under which conditions a component in a real solution tends to ideal behaviour. That is, if a component is the solvent, then ideal behaviour is achieved as the composition of the solution approaches pure solvent, i.e.

> as $x_i \rightarrow 1$, then $a_i \rightarrow x_i$ and $\gamma_i \rightarrow 1$

and therefore standard state III is employed. If a component is considered as a solute, then ideal behaviour is attained as the amount of solute tends to zero, i.e. the solution tends to infinite dilution, so that standard state IV is employed,

> as $x_i \rightarrow 0$, then $a_i \rightarrow x_i$ and $\gamma_i \rightarrow 1$

Alternatively, according to standard state V,

$$\text{as } m_i \rightarrow 0, \text{ then } a_i \rightarrow m_i \text{ and } \gamma_{mi} \rightarrow 1.$$

The properties of a solvent i in its standard state III have the same values as those of the pure substance at $101\,325\,N\,m^{-2}$ and the temperature under consideration, e.g. the standard partial molar volume, $\bar{V}_i{}^\circ$, is equal to v_i, the molar volume of pure i at $101\,325\,N\,m^{-2}$ and the temperature under consideration, and the standard partial molar enthalpy, $\bar{H}_i^\circ$, is equal to h_i, the molar enthalpy of pure i at $101\,325\,N\,m^{-2}$ and the temperature under consideration.

The intensive properties of a solute in its standard state IV have the same values as those of the solute at infinite dilution at $101\,325\,N\,m^{-2}$ and at the temperature under consideration, e.g. the standard partial molar volume is the same as the molar volume at infinite dilution and the standard partial molar enthalpy is the same as the molar enthalpy at infinite dilution. This is so for both solute standard states IV and V.

13.10 Use of the various standard states

For a gas solution, we would use standard state I for all the components when using Eq. (13.2) and standard state II when using Eq. (13.4). For a liquid solution, standard state III is always used for the solvent. If all the components are liquids at the same temperature and pressure as the solution, standard state III is used for all the components. If some of the components are gases or solids at the temperature and pressure of the solution, standard state IV is employed for these components (and III for the solvent). For aqueous solutions, standard state V may be used for gas or solid solutes.

We shall henceforth, unless otherwise stated, use the symbol $\ominus$ to indicate a standard state whether or not it is pressure-dependent, i.e. we shall use the symbol $\ominus$ where previously we used $\ominus$ or $\bullet$. Equation (13.13) will be written as

$$\mu_i = \mu_i^\ominus + RT \ln x_i \qquad (13.25)$$

it being understood that the standard chemical potential is pressure-dependent. The standard chemical potential, in general, need not be pressure-dependent; this depends on the standard state employed. Equation (13.8) becomes

$$\mu_i = \mu_i^\ominus + RT \ln a_i \qquad (13.26)$$

Equation (13.11) becomes

$$\mu_i = \mu_i^\ominus + RT \ln \frac{f_i}{f_i^\ominus} \qquad (13.27)$$

Equation (13.22) becomes

$$\mu_i = \mu_i^\ominus + RT \ln m_i \qquad (13.28)$$

13.11 Properties of solutions

For each component of a solution, the chemical potential is given at a particular temperature and pressure by Eq. (13.26),

$$\mu_i = \mu_i^\ominus + RT \ln a_i$$

the standard chemical potential, in addition to being temperature-dependent, possibly being pressure-dependent. The activity a_i is, of course, temperature-, pressure- and composition-dependent, $a_i = \gamma_i x_i$, the value of the activity (unlike fugacity) being dependent on the standard chosen for that component. If the solution were ideal, then $a_i = x_i$ and $\gamma_i = 1$.

Differentiating Eq. (13.26) with respect to pressure at constant temperature and composition, we have

$$\left(\frac{\partial \mu_i}{\partial P}\right)_{T,n} = \left(\frac{\partial \mu_i^\ominus}{\partial P}\right)_{T,n} + RT\left(\frac{\partial \ln a_i}{\partial P}\right)_{T,n}$$

Employing Eq. (10.32),

$$\boxed{\bar{V}_i = \bar{V}_i^\ominus + RT\left(\frac{\partial \ln a_i}{\partial P}\right)_{T,n}} \qquad (13.29)$$

where $\bar{V}_i$ is the partial molar volume of i in the solution, and the value of $\bar{V}_i^\ominus$ depends on the standard state chosen but must of course be independent of composition, since so is $\mu_i^\ominus$.

If the standard state chosen for i is based on the concept of ideal behaviour being achieved as the composition of the solution approaches that of pure i (i.e. standard state III) where $\gamma_i \to 1$ as $x_i \to 1$, then $\bar{V}_i^\ominus$ is equal to the molar volume of pure component i at the temperature and pressure of the solution: $\bar{V}_i^\ominus = v_i$. If the standard state chosen for i is based on ideal behaviour being achieved as the amount of solute i tends to zero (i.e. standard states I, IV and V) where $\gamma_i \to 1$ as $x_i \to 0$, then $\bar{V}_i^\ominus$ is equal to the partial molar volume of component i at infinite dilution at the temperature and pressure of the solution, i.e. the molar volume of component i in its hypothetical pure state.

IDEAL SOLUTIONS

For ideal solutions, $a_i = x_i$ and therefore Eq. (13.29) reduces to

$$\bar{V}_i = \bar{V}_i^\ominus \qquad (13.30)$$

Now according to our basic requirements, the value of $\bar{V}_i^\ominus$ depends on temperature and pressure only, i.e. whatever the standard state, $\bar{V}_i^\ominus$ is independent of composition. Therefore, from Eq. (13.30), the partial molar volume of component i in an ideal solution is independent of the composition of the solution. The volume of a solution is given by Eq. (10.18), i.e.

$$V = \Sigma_i n_i \bar{V}_i$$

Hence, for an ideal solution, the volume of the solution is independent of the composition of the solution (i.e. the molar volume of an ideal solution is independent of composition).

For a perfect solution, i.e. a solution that remains ideal over the complete range of composition (standard state III for all components), the partial molar volume of i is equal to the molar volume of i; $\bar{V}_i = v_i$. Hence, the volume of the solution is equal to the sum of the volumes of the components before mixing at the same temperature and pressure as the solution. That is, on preparing a perfect solution from its individual components, there is no change in volume on mixing.

PROPERTIES OF SOLUTIONS

Rearranging Eq. (13.26), we have

$$\frac{\mu_i}{T} = \frac{\mu_i^{\ominus}}{T} + R \ln a_i$$

Differentiating with respect to temperature at constant pressure and composition,

$$\left(\frac{\partial \mu_i/T}{\partial T}\right)_{P,n} = \left(\frac{\partial \mu_i^{\ominus}/T}{\partial T}\right)_{P,n} + R\left(\frac{\partial \ln a_i}{\partial T}\right)_{P,n}$$

Employing Eq. (10.36),

$$\boxed{\frac{\bar{H}_i}{T^2} = \frac{\bar{H}_i^{\ominus}}{T^2} - R\left(\frac{\partial \ln a_i}{\partial T}\right)_{P,n}} \qquad (13.31)$$

where $\bar{H}_i$ is the partial molar enthalpy of i in the solution and the value of $\bar{H}_i^{\ominus}$ depends on the standard state chosen but must be independent of composition as is $\mu_i^{\ominus}$.

As before, if the standard state is chosen on the basis that $\gamma_i \rightarrow 1$ as $x_i \rightarrow 1$, then $\bar{H}_i^{\ominus}$ is equal to the molar enthalpy of pure i at the same temperature and pressure as the solution, $\bar{H}_i^{\ominus} = h_i$. If the standard state is chosen on the basis that $\gamma_i \rightarrow 1$ as $x_i \rightarrow 0$, then $\bar{H}_i^{\ominus}$ is the partial molar enthalpy of component i at infinite dilution at the temperature and pressure of the solution.

IDEAL SOLUTIONS

For an *ideal solution*, $a_i = x_i$ and therefore Eq. (13.31) reduces to

$$\boxed{\bar{H}_i = \bar{H}_i^{\ominus}} \qquad (13.32)$$

Since $\bar{H}_i^\ominus$ must be independent of composition, the partial molar enthalpy of component i in an ideal solution must also be independent of composition. The enthalpy of the solution is given by Eq. (10.18),

$$H = \Sigma_i n_i \bar{H}_i$$

and therefore the enthalpy of an ideal solution is also independent of composition.

If the solution remains ideal over the complete range of composition, i.e. it is a perfect solution, then the partial molar enthalpy of i is equal to the molar enthalpy of pure component i at the temperature and pressure of the solution, $\bar{H}_i = h_i$. Hence, the enthalpy of a perfect solution is equal to the sum of the enthalpies of the components before mixing at the same temperature and pressure as the solution. That is, on preparing a perfect solution from its individual components, there is no change in enthalpy on mixing, which means, of course, that there is no heat of solution.

13.12   The temperature and pressure dependence of activity

Rearranging Eq. (13.29) gives the pressure dependence of activity,

$$\left(\frac{\partial \ln a_i}{\partial P}\right)_{T,\,n} = \frac{\bar{V}_i - \bar{V}_i^\ominus}{RT} \qquad (13.33)$$

Rearranging Eq. (13.31) gives the temperature dependence of activity,

$$\left(\frac{\partial \ln a_i}{\partial T}\right)_{P,\,n} = \frac{\bar{H}_i^\ominus - \bar{H}_i}{RT^2} \qquad (13.34)$$

Exactly similar equations hold for the activity coefficient. Since $a_i = \gamma_i x_i$, we have, at constant composition,

$$\left(\frac{\partial \ln a_i}{\partial T}\right)_{P,\,n} = \left(\frac{\partial \ln \gamma_i}{\partial T}\right)_{P,\,n}$$

Example   What is the approximate heat absorbed when 1 mole of copper is dissolved isothermally at 1330 K in a very large quantity of an ideal solution of copper in gold? The melting point of copper is 1356 K and its molar latent heat of fusion is 12·96 kJ mol^{-1}.

Since the solution is ideal, the heat absorbed on adding 1 mole of liquid copper to a large solution is zero. However, heat is involved in converting copper from the solid to the liquid phase at 1330 K. Since the melting point of copper is 1356 K, it is reasonable to say that this heat is approximately equal to the latent heat of fusion. Hence, the heat absorbed when 1 mole of copper is dissolved in the solution is approximately 12.96 kJ mol^{-1}.

13.13 Mixing functions

The enthalpy of mixing is defined as the difference between the enthalpy of the solution and the sum of the enthalpies of the pure components at the same temperature and pressure as the solution:

$$\Delta H_{\text{mix}} = H_{\text{solution}} - \Sigma_i H_i \qquad (13.35)$$

Other mixing functions may be defined similarly.

It is seen that the enthalpy of mixing for a perfect solution (one remaining ideal for all possible compositions) is zero:

$$\Delta H_{\text{mix}} = 0 \qquad (13.36)$$

It is logical that for ideal gas mixtures we should have $\Delta V_{\text{mix}} = 0$ and $\Delta H_{\text{mix}} = 0$, since Dalton's law may be considered as expressing that each ideal gas in a mixture disregards the others.

The total Gibbs function for a solution is given by

$$G = \Sigma_i n_i \mu_i \qquad (13.37)$$

so that $\qquad\qquad G = \Sigma_i n_i \mu_i^{\ominus} + RT \Sigma_i n_i \ln a_i \qquad (13.38)$

For an ideal solution, $a_i = x_i$, and if it is a perfect solution then $\mu_i^{\ominus}$ is the chemical potential of pure component i at the temperature and pressure of the solution. The chemical potential of a pure substance is the molar Gibbs function for the substance. Hence, the Gibbs function of mixing for a perfect solution is given by

$$\Delta G_{\text{mix}} = RT \Sigma_i n_i \ln x_i \qquad (13.39)$$

Multiplying and dividing by the total number of moles in the solution, $\Sigma_i n_i = N$,

$$\Delta G_{\text{mix}} = NRT \Sigma_i x_i \ln x_i \qquad (13.40)$$

Since the enthalpy of mixing for a perfect solution is zero, the entropy of mixing is given by

$$\Delta S_{mix} = -R \, \Sigma_i n_i \ln x_i \qquad (13.41)$$

since $\qquad T\Delta S_{mix} = \Delta H_{mix} - \Delta G_{mix}$

Or, multiplying and dividing by $\Sigma_i n_i = N$, we have

$$\Delta S_{mix} = -NR \, \Sigma_i x_i \ln x_i \qquad (13.42)$$

There is an increase in entropy on mixing since $x_i < 1$ and therefore $\ln x_i$ is negative. This indicates that the mixing is spontaneous. Similarly, ΔG_{mix} is negative.

For a binary perfect solution, Eqs (13.42), (13.36) and (13.40) apply. Using these equations, we may compute the values of the corresponding mixing functions for binary solutions of various compositions and plot the results as in Figure 13.3.

For example, for one mole of the binary ideal solution composed of 70% of A, let us calculate Δg_{mix} and Δs_{mix} at 298 K:

$$\Delta G_{mix} = NRT \left[x_A \ln x_A + (1 - x_A) \ln (1 - x_A) \right]$$

$$\Delta g_{mix} = 8 \cdot 314 \times 298 \left[0 \cdot 7 \ln 0 \cdot 7 + 0 \cdot 3 \ln 0 \cdot 3 \right] = -1513 \text{ J mol}^{-1}$$

$$\Delta S_{mix} = -\frac{\Delta g_{mix}}{T} = 5 \cdot 080 \text{ J mol}^{-1} \text{ K}^{-1}$$

It is interesting to note that the maximum entropy of mixing (or minimum Gibbs function of mixing) occurs for a binary perfect solution of composition $x_A = 0 \cdot 5$. This, of course, corresponds to the ideal solution having maximum disorder. Similarly, for a ternary perfect solution, the maximum entropy of mixing occurs for the composition $x_A = x_B = x_C = \frac{1}{3}$.

Equations (13.29) and (13.31) must be employed with Eq. (10.18) to determine the volume and enthalpy of mixing for non-ideal solutions, these quantities being non-zero. Figure 13.4 shows the entropy, enthalpy and Gibbs function of mixing for a non-ideal solution at various compositions.

13.14   Excess functions

The excess Gibbs function is defined as the difference between the Gibbs function for the solution and the Gibbs function for the

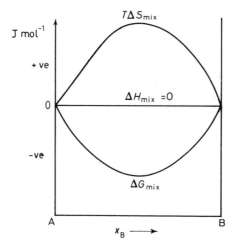

Figure 13.3. Variation of mixing functions with composition for a perfect solution

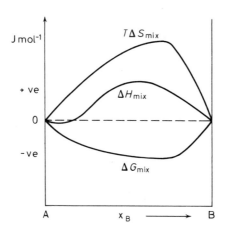

Figure 13.4. Variation of mixing functions with composition for a non-ideal solution

solution assumed ideal,

$$G_E = G_{\text{real solution}} - G_{\text{ideal solution}} \qquad (13.43)$$

Other excess functions may be defined similarly. These excess functions measure the deviation of a real from an ideal solution:

$$G_E = \Sigma_i n_i \mu_{i \text{ (real)}} - \Sigma_i n_i \mu_{i(\text{ideal})}$$

Employing Eq. (13.26) for the real and ideal solutions,

$$G_E = \Sigma_i n_i RT \ln a_i - \Sigma_i n_i RT \ln x_i$$

$$G_E = RT \Sigma_i n_i \ln \gamma_i \qquad (13.44)$$

Multiplying and dividing by the total number of moles, $\Sigma_i n_i = N$, we obtain

$$G_E = NRT \Sigma_i x_i \ln \gamma_i \qquad (13.45)$$

Equations for other excess functions involving the activity coefficient may be derived, but these are a little more complicated. For example, employing the Gibbs–Helmholtz equation, the excess enthalpy is found to be

$$H_E = -NRT^2 \Sigma_i x_i \left(\frac{\partial \ln \gamma_i}{\partial T} \right)_{P, n}$$

13.15   The Nernst distribution law

Consider a two-phase system consisting of two liquids which are immiscible and a component i that is present in both liquid layers, as in Figure 13.5. At equilibrium, the amounts of component i in the two phases are different, but the fugacity of i in both phases must be the same:

$$f_i = f_i'$$

Applying Eq. (13.17) to the two phases,

$$a_i f_i^{\ominus} = a_i' f_i'^{\ominus}$$

$$\frac{a_i}{a_i'} = \frac{f_i'^{\ominus}}{f_i^{\ominus}} = K \qquad (13.46)$$

where K is a constant which depends only on temperature and pressure and not on composition.

If each solution were ideal, then we would have

$$\frac{x_i}{x_i'} = K \qquad (13.47)$$

K is known as the partition coefficient. It is seen that the relative amounts of component i in the two phases is independent of composition.

Alternative proof Since, at equilibrium, the chemical potential of

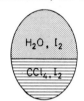

Figure 13.5.

i in both phases must be equal, we have $\mu_i = \mu_i'$, so that

$$\mu_i^\ominus + RT \ln a_i = \mu_i'^\ominus + RT \ln a_i'$$

$$\ln \frac{a_i}{a_i'} = \frac{\mu_i'^\ominus - \mu_i^\ominus}{RT} \qquad (13.48)$$

The right-hand side of this equation is constant at a given temperature and pressure, hence so is a_i/a_i'.

13.16   A pure substance in equilibrium with its solution

We have seen in Section 12.20 that there is no inherent difference between solubility and freezing-point curves. They are both determined by an equilibrium between a pure solid i and its solution, for which we have $\mu_i' = \mu_i$, the prime denoting the pure solid. The chemical potential of i in solution is given by Eq. (13.26), so that we have

$$\mu_i' = \mu_i^\ominus + RT \ln a_i$$

Rearranging,

$$\ln a_i = \frac{\mu_i'}{RT} - \frac{\mu_i^\ominus}{RT}$$

Differentiating with respect to temperature at constant pressure, and employing Eq. (10.36),

$$\left(\frac{\partial \ln a_i}{\partial T} \right)_P = \frac{\bar{H}_i^\ominus - h_i'}{RT^2} \qquad (13.49)$$

where h'_i is the molar enthalpy of the pure solid and $\bar{H}_i^{\ominus}$ is the standard partial molar enthalpy of i in the solution and is dependent on the choice of standard state.

Equation (13.49) *applies to any system in which there is equilibrium between a pure substance and its solution*, e.g. *pure solid in equilibrium with its solution* or *pure gas in equilibrium with a liquid solution containing the gas*. If we are considering the depression of freezing point, the pure solid is the solvent in equilibrium with the solution. The standard state of i is defined as for a solvent, i.e. so that as $x_i \to 1$, $\gamma_i \to 1$. Hence, $\bar{H}_i^{\ominus}$ is the molar enthalpy of pure liquid i at the temperature and pressure under consideration and $\bar{H}_i^{\ominus} - h'_i$ is the latent heat of melting L_i at the temperature T and pressure P. We may rewrite Eq. (13.49) as

$$\left(\frac{\partial \ln a_i}{\partial T} \right)_P = \frac{L_i}{RT^2} \qquad (13.50)$$

If we are considering solubilities, then i is a solute and its standard state is defined so that as $x_i \to 0$, $\gamma_i \to 1$. Hence, $\bar{H}_i^{\ominus}$ is the molar enthalpy of component i at infinite dilution. In this case, L_i is the differential heat of solution, i.e. the heat evolved when 1 mole of solute i dissolves at the given temperature and pressure in a large quantity of solution whose composition remains unaltered. Equation (13.50) applies to both solubility or freezing point phenomena; it shows how the solubility depends on temperature. If the solution is ideal, then $a_i = x_i$ and, hence,

$$\left(\frac{\partial \ln x_i}{\partial T} \right)_P = \frac{L_i}{RT^2} \qquad (13.51)$$

13.17 Depression of freezing point

Experiments dealing with this phenomenon usually involve dilute solutions, so that the depression of the freezing point is only a few degrees. Hence, we may regard L_i to be independent of temperature and approximately equal to the latent heat of melting L_m of pure component i. Also, the solution may be assumed to be ideal.

Integrating Eq. (13.51) from the state of pure component i to that of solution with mole fraction x_i at constant pressure, we have

$$\int_1^{x_i} \mathrm{d} \ln x_i = \frac{L_m}{R} \int_{T_m}^{T} \frac{\mathrm{d}T}{T^2}$$

where T_m and T are the melting points of pure component i and the solution, respectively.

$$\ln x_i = -\frac{L_m}{R}\left(\frac{1}{T} - \frac{1}{T_m}\right) = -\frac{L_m}{R}\left(\frac{T_m - T}{T T_m}\right)$$

Now if more than one solute is added, $x_i = 1 - \Sigma_j x_j$ where $\Sigma_j x_j$ is the sum of the mole fractions of all the solutes. $T_m - T = \Delta T$ is the depression of the freezing point. If we assume that $T \doteq T_m$, i.e. the depression is only of the order of a few degrees, then we may write

$$\ln (1 - \Sigma_j x_j) = -\frac{L_m \Delta T}{R T_m^2}$$

The logarithmic term may be expanded in the form of a series since, in general,

$$\ln (1 - b) = -b - \frac{b^2}{2} - \frac{b^3}{3} - \cdots \qquad (13.52)$$

where b is equivalent to our $\Sigma_j x_j$. If we assume that $\Sigma_j x_j$ is small compared with unity, the power terms may be neglected, hence

$$\boxed{\Sigma_j x_j = \frac{L_m \Delta T}{R T_m^2}} \qquad (13.53)$$

For a binary solution, the summation sign would be omitted. On measuring the depression of the freezing point of a binary solution, the mole fraction of the solute may be calculated and, hence, if the weights of the solvent and solute are known, the molecular weight of the solute may be determined.

Equation (13.53) is rather remarkable since it shows that the depression of the freezing point depends on the mole fraction of the solute but not on the *nature* of the solute. Properties which depend on the quantity of a substance without regard to its nature are known as *colligative* properties. Another example of a colligative property is the partial pressure of an ideal gas in a mixture, since this depends only on the amount of the gas in the mixture at a given temperature and pressure.

A more exact equation for the depression of freezing point may be obtained on taking account of the temperature dependence of L_i. Differentiating L_i with respect to temperature at constant pressure,

$$\left(\frac{\partial L_i}{\partial T}\right)_P = \left(\frac{\partial H_i^\ominus}{\partial T}\right)_P - \left(\frac{\partial h_i'}{\partial T}\right)_P = \bar{C}_P^\ominus - c_P' = \Delta c_P \qquad (13.54)$$

where $\bar{C}_P^{\ominus}$ is the partial molar heat capacity of solvent i in the standard state, which is equal to the molar heat capacity of pure liquid i for an ideal solution, and c_P' is the molar heat capacity of pure solid i.

Integrating from the state of pure liquid i to that of solution in equilibrium with pure solid i at constant pressure, assuming that Δc_P is independent of temperature over the range of the freezing-point depression,

$$\int_{L_{\mathrm{m}}}^{L_i} \mathrm{d}L_i = \Delta c_P \int_{T_{\mathrm{m}}}^{T} \mathrm{d}T$$

$$L_i - L_{\mathrm{m}} = \Delta c_P (T - T_{\mathrm{m}}) \tag{13.55}$$

Hence, substituting into Eq. (13.51),

$$\left(\frac{\partial \ln x_i}{\partial T}\right)_P = \frac{L_{\mathrm{m}} + \Delta c_P (T - T_{\mathrm{m}})}{RT^2}$$

Integrating from $x_i = 1$ to x_i at constant pressure,

$$\int_{x_i = 1}^{x_i} \mathrm{d} \ln x_i = \frac{L_{\mathrm{m}} - \Delta c_P T_{\mathrm{m}}}{R} \int_{T_{\mathrm{m}}}^{T} \frac{\mathrm{d}T}{T^2} + \frac{\Delta c_P}{R} \int_{T_{\mathrm{m}}}^{T} \frac{\mathrm{d}T}{T}$$

$$\ln x_i = \frac{L_{\mathrm{m}} - \Delta c_P T_{\mathrm{m}}}{R} \left(\frac{1}{T_{\mathrm{m}}} - \frac{1}{T}\right) + \frac{\Delta c_P}{R} \ln \frac{T}{T_{\mathrm{m}}} \tag{13.56}$$

13.18   Pure vapour in equilibrium with a solution: elevation of boiling point

Only if the solutes of a solution are involatile can a pure vapour be in equilibrium with a solution, i.e. if one or more involatile solutes are dissolved in a volatile solvent. Hence, for a liquid solution, Eq. (13.51) gives the dependence of the boiling point at a given pressure on the composition of the ideal solution. Equations (13.57) and (13.58), similar to Eqs (13.53) and (13.56), respectively, may be derived by assuming that over the range considered the latent heat of vaporisation of the solvent from solution is constant or temperature-dependent, respectively:

$$\Sigma_j x_j = \frac{L_{\mathrm{B}} \Delta T}{RT_{\mathrm{B}}^2} \tag{13.57}$$

$$\ln x_i = \frac{L_{\mathrm{B}} - \Delta c_P T_{\mathrm{B}}}{R} \left(\frac{1}{T_{\mathrm{B}}} - \frac{1}{T}\right) + \frac{\Delta c_P}{R} \ln \frac{T}{T_{\mathrm{B}}} \tag{13.58}$$

where L_B is the latent heat of vaporisation of the pure solvent; T_B and T are the boiling points of the pure solvent and solution, respectively;

ΔC_P is the difference between the molar heat capacity of the solvent vapour and that of the pure solvent liquid; and ΔT is the elevation of the boiling point of the solvent.

It can be seen from Eq. (13.57) that the elevation of the boiling point depends only on the amount of the solute and not on its properties. Hence, the elevation of boiling point is a colligative property.

13.19 Osmotic pressure

Osmosis is the spontaneous flow of solvent into solution when the solvent and solution are separated by a semipermeable membrane. The semipermeable membrane allows the passage of solvent molecules but not those of the solute. There are many types of semipermeable membrane. For example, for water, a porous pot impregnated with cupric ferrocyanide $Cu_2Fe(CN)_6$, and for solvents such as pyridine or acetone, vulcanised rubber may be used.

Osmotic pressure is the excess pressure which must be applied to a solution in order just to prevent osmosis. That is, the excess pressure which must be applied to attain equilibrium. If the pressure is greater than this, there will be a net diffusion of solvent from the solution to the pure solvent. Osmotic pressure is a property of a solution and is in no way dependent on the nature of the membrane.

Since in osmosis there is a mass transfer, the chemical potential of the solvent in the solution must be less than that of the pure solvent,

$$\mu_{solvent}^{solution} < \mu'_{solvent}$$

where the prime denotes the pure phase. (In other words, the Gibbs function of mixing is negative.)

Consider a solvent separated from its solution by a semipermeable membrane, the system having the same temperature throughout. Let the pressure on the solution and solvent at equilibrium be P and P', respectively. Then the osmotic pressure is given by

$$\pi = P - P' \tag{13.59}$$

At equilibrium, the chemical potential of the solvent on either side of the membrane must be the same;

$$\mu_{solvent}^{solution} = \mu'_{solvent}$$

Substituting for the chemical potential of the solvent in the solution, using Eq. (13.26),

$$\mu_i^{\ominus} + RT \ln a_i = \mu_i'$$

$$\mu_i' - \mu_i^{\ominus} = RT \ln a_i$$

Now $\mu_i^{\ominus}$ is equal to the chemical potential of pure component i at

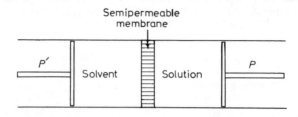

Figure 13.6. Osmotic pressure

the pressure P and temperature T. Hence, employing Eq. (10.32) at a given temperature, we have

$$\int_P^{P'} \bar{V}_i \, dP = RT \ln a_i \qquad (13.60)$$

where $\bar{V}_i$ is the partial molar volume of i.

If we make the assumption that $\bar{V}_i$ is independent of pressure, we obtain

$$\bar{V}_i(P' - P) = RT \ln a_i$$

or, $$-\bar{V}_i \pi = RT \ln a_i \qquad (13.61)$$

If the solution is ideal, then $a_i = x_i$, so that

$$-\bar{V}_i \pi = RT \ln x_i = RT \ln(1 - \Sigma_j x_j)$$

where $\Sigma_j x_j$ is the sum of the mole fractions of the solutes. Assuming that $\Sigma_j x_j$ is small, i.e. that we are dealing with a very dilute solution, we have, from Eq. (13.52),

$$\ln(1 - \Sigma_j x_j) \doteqdot -\Sigma_j x_j$$

Hence, $$\bar{V}_i \pi = \Sigma_j x_j \cdot RT$$

Now $$\Sigma_j x_j = \frac{\Sigma_j n_j}{N}$$

where $\Sigma_j n_j$ is the total number of moles of solute, and N is the total number of moles of the solution ($N = n_i + \Sigma_j n_j$). In a very dilute solution, N is approximately equal to the number of moles of solvent

n_i in the solution. Hence,

$$\pi \, n_i \, \bar{V}_i = \Sigma_j n_j \, RT$$

For an ideal solution, $n_i \bar{V}_i$ is the volume of the solution occupied by the solvent. Since the solution is very dilute, $n_i \, \bar{V}_i$ is approximately equal to the total volume of the solution, V:

$$\boxed{\pi V = \Sigma_j n_j \, RT} \qquad (13.62)$$

It is interesting to compare this equation with the ideal gas equation of state. It can be seen that osmotic pressure is a colligative property.

The osmotic pressures of even very dilute solutions are extremely large. For example, dilute solutions of sugar in water have osmotic pressures of $2\cdot5$ MN m^{-2} – 10 MN m^{-2} (approximately 25–100 atm) at room temperature. Hence, $\bar{V}_i$ cannot strictly be considered, in Eq. (13.60), a constant over such large ranges of pressure.

13.20 Determination of molecular weight

The colligative properties elevation of boiling point, depression of freezing point and osmotic pressure may be employed to determine the molecular weight of a solute. Equations (13.53), (13.57) and (13.62) have been derived for ideal solutions. Association or dissociation of solute molecules in solution must be allowed for in certain circumstances, e.g. the elevation of boiling point gives double the molecular weight of benzoic acid in a benzene solution, thus showing that the acid exists as a dimer in the solution. If hydrogen bonding is present, different values of the molecular weight are obtained at different concentrations. Also, incorrect values for the molecular weight are obtained if ionisation occurs.

13.21 Significance of the activity coefficient

Consider a binary solution showing negative deviations from Raoult's law, and for simplicity let the vapour over the solution behave ideally. If the solution is such that for component B it is convenient to choose standard state III, where $\gamma_B \to 1$ as $x_B \to 1$, then the vapour pressure of B (assuming the vapour to be ideal) is given by

$$P_B = \gamma_B x_B P_B^{\ominus} \qquad (13.63)$$

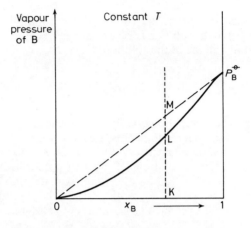

Figure 13.7.

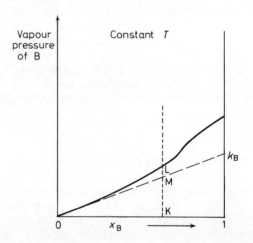

Figure 13.8.

For a solution of composition K (Figure 13.7), the vapour pressure of B, P_B, is given by KL. The vapour pressure predicted by Raoult's law, $x_B P_B^\ominus$, is given by KM. Hence, from Eq. (13.63), the activity coefficient is given by

$$\gamma_B = \frac{KL}{KM}$$

If however we choose the standard state IV where $\gamma_B \to 1$ as $x_B \to 0$, then the vapour pressure of B is given by

$$P_B = \gamma_B x_B K_B$$

where K_B is the Henry's law constant, i.e. the slope of the tangent to the curve at infinite dilution.

The vapour pressure of a binary solution showing negative deviations from Raoult's law (Figure 13.8), of composition K, is given by KL, and the vapour pressure predicted by Henry's law, $x_B K_B$, is given by KM. Therefore

$$\gamma_B = \frac{KL}{KM}$$

The activity coefficients may thus be obtained from direct measurements of vapour pressure.

13.22 Determination of activity coefficients

VAPOUR PRESSURE MEASUREMENTS

This method has already been discussed in the previous section and may be used for a component which has an appreciable vapour pressure. It is normally used to determine the activity coefficients of solvents. In general, the vapour pressures P_B and $P_B^\ominus$ may be replaced by the corresponding fugacities.

DEPRESSION OF FREEZING POINT

From Section 13.16, we have

$$\left(\frac{\partial \ln a_i}{\partial T} \right)_P = \frac{L_m}{RT^2}$$

where a_i and L_m are the activity and the latent heat of melting of the solvent. It can therefore be seen that, to a first approximation,

$$\ln \gamma_i x_i = -\frac{L_m \Delta T}{R T_m^2}$$

where ΔT is the depression of the freezing point. and T_m is the freezing point of the solvent. Hence, the activity coefficient of the solvent may be found.

Alternatively, if the temperature variation of the latent heat is taken into account, we have

$$\ln \gamma_i x_i = \frac{L_m - \Delta c_P T_m}{R}\left(\frac{1}{T_m} - \frac{1}{T}\right) + \frac{\Delta c_P}{R}\ln\frac{T}{T_m}$$

where T_m and T are the freezing points of the pure solvent and the solution. This equation enables the activity of the solvent to be found.

OSMOTIC PRESSURE

For non-ideal solutions, we have derived Eq. (13.61). The activity of the solvent may be determined from

$$\ln a_i = -\frac{\pi \bar{V}_i}{RT}$$

NERNST DISTRIBUTION

We have shown that if a solute is distributed between two solvents, α and β, then, from Eq. (13.46),

$$\frac{a^\alpha}{a^\beta} = K$$

where a^α and a^β are the activities of the solute in solvents α and β, and K is a constant at a given temperature and pressure. The constant K may be obtained by extrapolating x^α/x^β to infinite dilution. Hence, if the activity coefficient of the solute in solvent α is known by some other method, then γ^β may be found.

THE GIBBS–DUHEM EQUATION

This method is used for binary solutions if the variation of the activity of one component with composition is known. At constant temperature and pressure, the Gibbs–Duhem equation for such solutions may be written as

$$x_A \, d\mu_A + x_B \, d\mu_B = 0$$

Using the differential form of the equations

$$\mu_A = \mu_A^\ominus + RT \ln a_A$$

and
$$\mu_B = \mu_B^\ominus + RT \ln a_B$$

we have,

$$x_A \, d \ln a_A + x_B \, d \ln a_B = 0$$

That is, the activity of one component is related to the activity of the other component. Rearranging,

$$d \ln a_A = -\frac{x_B}{1 - x_B} d \ln a_B$$

Integrating between $x_B = 1$ and $x_B = x_B'$, we have

$$\int_{a_A=1}^{a_A} d \ln a_A = -\int_{x_B=1}^{x_B'} \frac{x_B}{1 - x_B} d \ln a_B \qquad (13.64)$$

where, from standard state IV, $a_A \to x_A$ as $x_A \to 0$. Hence, if the variation of a_B with x_B is known, then the integral may be evaluated from the area under the graph of $x_B/(1 - x_B)$ against $\ln a_B$. This integration presents a problem, since, as $x_B \to 1$, $x_B/(1 - x_B) \to \infty$, making it difficult to estimate the area. This may be overcome by integrating Eq. (13.64) not from $x_B = 1$ but from a value of x_A at which the solvent approximately obeys Raoult's law, i.e. so that $a_B = x_B$, when $a_A = x_A$ from Henry's law. The integral is then being evaluated between x_B and x_B'.

Problems

$$R = 8{\cdot}314 \, \text{J K}^{-1} \, \text{mol}^{-1}$$

1. The osmotic pressure of a solution containing $1{\cdot}345 \times 10^{-3}$ kg in 10^{-4} m^3 of water at 298 K is $971{\cdot}5$ N m^{-2}. What is the molecular weight of the solute?
Answer 34 150

2. The osmotic pressure of a polymer solution of concentration $4 \, \text{kg m}^{-3}$ is $64{\cdot}8$ N m^{-2} at 300 K. Calculate the mean molecular weight of the solution.
Answer $1{\cdot}5 \times 10^5$

3. The boiling point of benzene is 353·2 K and its molar heat of vaporisation is 31·35 kJ mol^{-1}. If 0·1375 kg of diphenyl is added to 1 kg of benzene, calculate the elevation of the boiling point.
Answer　1·27 K

4. Calculate the elevation of boiling point of an aqueous solution containing a solute of mole fraction 0·01. The molar latent heat of vaporisation of water is 40·55 kJ mol^{-1}.
Answer　0·29 K

5. If 0·05 kg of sucrose, $C_{12}H_{22}O_{11}$, is added to 1 kg of water, calculate the elevation of boiling point. The latent heat of vaporisation of water is 2·254 MJ kg^{-1}.
Answer　0·075 K

6. The melting point of pure naphthalene is 353·3 K and its latent heat is 148 300 J kg^{-1}. If a sample is 99·9 mole per cent pure, calculate the melting point of the sample.
Answer　352·7 K

7. Define (a) ideal solution, (b) perfect solution, (c) activity. Show that the enthalpy of mixing for an ideal solution is zero. Discuss how activity coefficients may be determined.

8. Calculate the entropy of mixing for solutions of 40 and 80 mole per cent of *o*-xylene in *m*-xylene, assuming perfect solutions are formed. The entropies of *o*- and *m*-xylene are 247·7 and 251·9 J K^{-1} mol^{-1}, respectively.
Answer　260·6 J K^{-1} mol^{-1} and 252·7 J K^{-1} mol^{-1}

9. The latent heat of fusion and the melting point of pure dimethylamine are 5·936 kJ mol^{-1} and 180·97 K. If a sample melts at 180·87 K, calculate the mole per cent of impurity in the sample.

10. Calculate the solubility at 78 K of methane in liquid nitrogen, assuming that an ideal solution is formed. The melting point of methane is 90·5 K and its molar latent heat of fusion is 970 J mol^{-1}.
Answer　Mole fraction = 0·815

14
Reaction Equilibria

14.1 van't Hoff isotherm

Consider the general reaction

$$aA + bB \longrightarrow cC + dD \qquad (14.1)$$

where A, B are reactants, C, D are products and a, b, c, d are the stoichiometric coefficients of the respective substances. At some stage of the reaction, let dn_A moles of A and dn_B moles of B proceed to react to form dn_C and dn_D moles of C and D, respectively. It follows (from the law of definite proportions) that

$$dn_A = -a \cdot d\alpha, \qquad dn_B = -b \cdot d\alpha$$
$$dn_C = c \cdot d\alpha \quad \text{and} \quad dn_D = d \cdot d\alpha \qquad (14.2)$$

where $d\alpha$ is a proportionality constant and α is a measure of the extent or progress of the reaction. Note that the velocity of the reaction at time t is $d\alpha/dt$. By integrating each of these equations from the initial state of the reaction ($\alpha = 0$) to a particular instant during the reaction, the amount of a species in the reaction vessel may be found, e.g. $n_A - a\alpha$ and $n_C + c\alpha$, where n_A and n_C are the initial amounts of the species A and C.

From Eq. (10.10) the change in the Gibbs function for the infinitesimal change in the reaction at constant temperature and pressure is

$$dG = \Sigma_i \mu_i \, dn_i$$

Employing Eq. (14.2),

$$dG = c\mu_C d\alpha + d\mu_D d\alpha - a\mu_A d\alpha - b\mu_B \, d\alpha$$

or,
$$dG = \Delta G \, d\alpha \qquad (14.3)$$

where
$$\Delta G = c\mu_C + d\mu_D - a\mu_A - b\mu_B \qquad (14.4)$$

285

and is the change in the Gibbs function for the system when c and d moles of species C and D are formed from a and b moles of reactants A and B under the particular conditions of the reaction at a given instant during its progress.

If the reaction has reached equilibrium, then $dG = 0$. Now $d\alpha$ is not zero, since it represents an arbitrary change in the extent of the reaction; therefore

$$\boxed{\Delta G = 0} \qquad (14.5)$$

This equation is of extreme importance. It may be regarded as a general condition for chemical equilibrium which applies whether the reaction is homogeneous or heterogeneous and whether the reactants and products are liquids, gases or solids.

The chemical potential of a species i is, in general, given by

$$\mu_i = \mu_i^\ominus + RT \ln a_i$$

Employing this for each of the species to calculate ΔG, we obtain

$$c\mu_C + d\mu_D - a\mu_A - b\mu_B = c\mu_C^\ominus + d\mu_D^\ominus - a\mu_A^\ominus - b\mu_B^\ominus \\ + cRT \ln a_C + dRT \ln a_D - aRT \ln a_A - bRT \ln a_B$$

i.e.

$$\boxed{\Delta G = \Delta G^\ominus + RT \ln \frac{a_C^c a_D^d}{a_A^a a_B^b}} \qquad (14.6)$$

where $\Delta G^\ominus$ is defined similarly to ΔG, i.e.

$$\Delta G^\ominus = c\mu_C^\ominus + d\mu_D^\ominus - a\mu_A^\ominus - b\mu_B^\ominus \qquad (14.7)$$

and is, of course, the change in the standard Gibbs function.

$\Delta G^\ominus$ is, therefore, constant at a given temperature and pressure. At equilibrium, $\Delta G = 0$; hence,

$$-\Delta G^\ominus = RT \ln \left\{ \frac{a_C^c a_D^d}{a_A^a a_B^b} \right\}_{eq} \qquad (14.8)$$

where the eq indicates that the activities are those at equilibrium and not at some stage during the reaction.

At a given temperature and pressure, $\Delta G^\ominus$, R and T are constants, and therefore

$$\left\{ \frac{a_C^c a_D^d}{a_A^a a_B^b} \right\}_{eq}$$

must also be a constant. This constant is known as the equilibrium constant, K_{eq}:

$$K_{eq} = \frac{a_C^c a_D^d}{a_A^a a_B^b} \qquad (14.9)$$

it being understood that the activities referred to are those at equilibrium. As a consequence of the dependencies of ΔG°, K_{eq} is independent of composition but dependent on temperature, and possibly pressure also.

We have thus proved thermodynamically what chemists know as the law of mass action.

For an ideal system, taking the activities as being based on mole fraction, we have

$$K_{eqx} = \frac{x_C^c x_D^d}{x_A^a x_B^b}$$

or, if based on molality,

$$K_{eqm} = \frac{m_C^c m_D^d}{m_A^a m_B^b}$$

Obviously, in general, $K_{eqx} \neq K_{eqm}$

Equation (14.8) may be written

$$-\Delta G^\circ = RT \ln K_{eq}$$

This very important equation is known as the *van't Hoff isotherm* and may be written in the equivalent form

$$K_{eq} = \exp\left(-\frac{\Delta G^\circ}{RT}\right)$$

Substituting for ΔG° in Eq. (14.6), we have

$$\Delta G = -RT \ln K_{eq} + RT \ln Q \qquad (14.11)$$

where

$$Q = \frac{a_C^c a_D^d}{a_A^a a_B^b} \qquad (14.12)$$

and is known as the reaction quotient. Q is not equal to the equilibrium constant except when the reaction is at equilibrium, since the activities in Eq. (14.12) are the activities of species at a point in the progress of the reaction under consideration.

Equation (14.11) may also be written

$$\Delta G = RT \ln \frac{Q}{K_{eq}}$$ (14.13)

This equation is known as the *reaction isotherm*.

14.2   Alternative (condensed) approach to reaction equilibria

It is *convenient* to write a chemical reaction

$$\text{reactants} \rightleftharpoons \text{products}$$

as

$$0 \quad \rightleftharpoons \text{products} - \text{reactants}$$

That is, in the general form

$$0 \rightleftharpoons \Sigma_i \, v_i \, C_i$$ (14.14)

where the v_i are the stoichiometric coefficients of the constituents C_i. For products, v_i is positive and for reactants, negative, e.g. in the reaction

$$N_2 + 3H_2 \rightleftharpoons 2NH_3$$

we have $v_{NH_3} = 2$, $v_{H_2} = -3$ and $v_{N_2} = -1$.

Consider dn_1 moles of C_1 reacting with dn_2 moles of $C_2 \ldots$, etc., to form dn_1 moles of C_i, etc. It follows (from the law of definite proportions) that

$$\frac{dn_1}{v_1} = \frac{dn_2}{v_2} = \cdots = \frac{dn_i}{v_1} = d\alpha$$ (14.15)

where α is a measure of the progress (i.e. the extent) of the reaction. For equilibrium at a given temperature and pressure the change in the Gibbs function is zero:

$$dG = \Sigma_i \, v_i \, \mu_i \, d\alpha = 0$$

Now $d\alpha$ is not zero, since it represents an arbitrary change in the progress of the reaction, so we must have

$$\Sigma_i \, v_i \, \mu_i = 0$$ (14.16)

or
$$\boxed{\Delta G = 0} \qquad (14.17)$$

This equation is the general condition for chemical equilibrium and applies whether the reactants and products are liquids, solids or gases, and whether the reaction is homogeneous or heterogeneous.

The chemical potential of a species i is given by

$$\mu_i = \mu_i^\ominus + RT \ln a_i$$

Substituting into Eq. (14.16),

$$0 = \Sigma_i \, v_i \, \mu_i^\ominus + \Sigma_i \, v_i \, RT \ln a_i$$

Therefore

$$-\Sigma_i \, v_i \, \mu_i^\ominus = RT \ln \Pi_i \, a_i^{v_i} \qquad (14.18)$$

where Π_i means the product over i terms.

Since the left-hand side is constant at a given temperature and pressure, $\Pi_i a_i^{v_i}$ must also be constant. It is, in fact, known as the equilibrium constant, K_{eq}. Hence,

$$\boxed{-\Delta G^\ominus = RT \ln K_{eq}} \qquad (14.19)$$

where

$$\Delta G^\ominus = \Sigma_i \, v_i \, \mu_i^\ominus \qquad (14.20)$$

and is the change in the standard Gibbs function when stoichiometric quantities of products are formed, reactants and products being in their standard states.

14.3   Equilibrium position

We have already discussed in Section 8.8 (and see also Section 8.9) how functions of state change as a closed system approaches equilibrium. Since the most common and easily produced conditions for a chemical reaction in a laboratory are either constant T and P or constant V and T, we are most interested in the properties G and A of a system (contents of the reaction vessel).

For example, for a reaction carried out at constant T and P, as equilibrium is approached the Gibbs function G of the system tends to a minimum. Consider the reaction

$$aA + bB = cC + dD$$

In Figure 14.1, G is plotted against composition. Initially, when only stoichiometric proportions of A and B are in the reaction vessel, the Gibbs function of the system is

$$G_{A,B}^\ominus = aG_A^\ominus + bG_B^\ominus$$

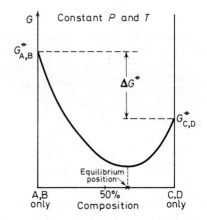

Figure 14.1. Variation in the Gibbs function during the reaction $a\mathrm{A}+b\mathrm{B} = c\mathrm{C}+d\mathrm{D}$

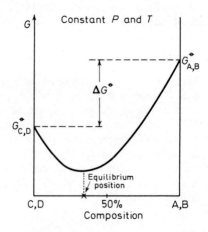

Figure 14.2. Variation in the Gibbs function during the reaction $c\mathrm{C}+d\mathrm{D} = a\mathrm{A}+b\mathrm{B}$

As the reaction proceeds spontaneously, the Gibbs function decreases and is a minimum at equilibrium. Any reaction proceeds spontaneously if the Gibbs function of the system decreases. If initially we placed only C and D in the reaction vessel, the reaction would still proceed spontaneously to the same equilibrium position. It is seen that at equilibrium the contents of the reaction vessel are rich in C and D. If we wish to produce more of one component, say C, all that need be done is to remove this component (or D, in this case) from the vessel. The equilibrium position, which is determined by the minimum in G, would not, in fact, be altered.

The standard Gibbs function change is

$$\Delta G^{\ominus} = G^{\ominus}_{\text{products}} - G^{\ominus}_{\text{reactants}} (= cG^{\ominus}_{C} + dG^{\ominus}_{D} - aG^{\ominus}_{A} - bG^{\ominus}_{B})$$

and this may be calculated for any reaction from tables. If $\Delta G^{\ominus}$ is a negative quantity, then the equilibrium position lies closer to the products than the reactants (see Figure 14.1). If $\Delta G^{\ominus}$ is positive, the equilibrium position lies near to the reactants and the reaction is not favoured, for example, as in Figure 14.2, which shows the reverse of the reaction considered above, i.e. $cC + dD = aA + bB$.

The fact that for a reaction $\Delta G^{\ominus}$ is positive does not mean that the reaction will not proceed. What it does imply is that the equilibrium lies near the reactants.

The larger (positive) $\Delta G^{\ominus}$, the closer the equilibrium position will be toward the reactants. If $\Delta G^{\ominus}$ is zero, the reactants and products are at equilibrium. Even though $\Delta G^{\ominus}$ may be calculated to be negative for a given reaction, this does not mean that the reaction will proceed with any speed. Thermodynamics does not give information on the velocity of processes. For example, hydrogen and oxygen may be kept in a container for a very long time before water may be detected, whereas if a catalyst is added, the reaction occurs with explosive violence and the equilibrium position is achieved very quickly.

The affinity of a reaction — that is, the tendency of the reaction to proceed — may be equated to the quantity $-\Delta G$ (i.e. $-\Sigma_i v_i \mu_i$). At equilibrium the affinity is zero. For a reaction to be spontaneous, the affinity must be a positive quantity.

Usually, when chemists record the standard Gibbs functions for compounds, they consider that (1) the standard state is the state of the substance at $101\ 325\ N\ m^{-2}$ and 298 K, (2) by convention the standard Gibbs function of the most stable form of an element is zero, and (3) the standard Gibbs function of a compound is the change in the Gibbs function for the reaction by which the compound is formed from its elements, all substances being in their standard states.

As may be seen, point (2) is of no value where nuclear reactions are concerned. The concepts above are not unlike those considered for standard enthalpies.

Example   Calculate the equilibrium constant for the reaction

$$CO(g) + H_2O(g) = H_2(g) + CO_2(g)$$

at 101 325 N m^{-2} and 298 K. The standard Gibbs functions $G^\ominus(298\,\text{K})$ for CO(g), H$_2$O(g) and CO$_2$(g) are $-137\cdot05$, $-228\cdot25$ and -393.94 kJ mol^{-1}.

The standard Gibbs function change, $\Delta G^\ominus$. for the reaction is

$$-393\cdot94 + 0 - (-228\cdot25 + (-137\cdot05)) = 27\cdot64\;\text{kJ mol}^{-1}$$

Hence, $$-\ln K_{P/P^\ominus} = \frac{27640}{298 \times 8\cdot314}$$

so that $$K_{P/P^\ominus} = 0\cdot082\,19$$

14.4   Homogeneous ideal gas reactions

Consider the reaction given by Eq. (14.1) or (14.14) as involving ideal gases only. The chemical potential of component i is given by

$$\mu_i = \mu_i^\ominus + RT \ln \frac{P_i}{P_i^\ominus}$$

In this equation, the *standard state I* is used, thus making $\mu_i^\ominus$ *dependent only on the temperature* and nature of the gas i. By following a similar derivation to that of Eq. (14.8), or (14.18), we obtain

$$\boxed{-\Sigma_i \, \nu_i \, \mu_i^\ominus = RT \ln K_{P/P^\ominus}} \qquad (14.21)$$

where the equilibrium constant of the ideal gas reaction in terms of partial pressures is

$$K_{P/P^\ominus} = \frac{(P_C/P_C^\ominus)^c(P_D/P_D^\ominus)^d}{(P_A/P_A^\ominus)^a(P_B/P_B^\ominus)^b} \quad \text{or} \quad K_{P/P^\ominus} = \prod_i \left(\frac{P_i}{P_i^\ominus}\right)^{\nu_i}$$

Hence, it is seen from Eq. (14.21) that $K_{P/P^\ominus}$ is independent of pressure.

If each of the gases have the same standard state (I), then

$$P_A^\ominus = P_B^\ominus = P_C^\ominus = P_D^\ominus = P^\ominus$$

Hence, $\quad K_{P/P^\ominus} = \dfrac{P_C^c P_D^d}{P_A^a P_B^b} \cdot (P^\ominus)^{-c-d+a+b} \quad$ or $\quad K_{P/P^\ominus} = (P^\ominus)^{-\Sigma v_i} \prod_i P_i^{v_i}$

i.e. $\quad K_{P/P^\ominus} = \dfrac{P_C^c P_D^d}{P_A^a P_B^b} \cdot (P^\ominus)^{-\Delta n} \quad$ or $\quad K_{P/P^\ominus} = (P^\ominus)^{-\Delta n} \prod_i P_i^{v_i}$

where $\Delta n = c + d - (a + b)$ or, more generally, $\Delta n = \Sigma_i v_i$.
We may write Eq. (14.21) as

$$\boxed{-\Delta G^\ominus = RT \ln K_{P/P^\ominus}} \tag{14.22}$$

Alternatively, if measuring mole fraction in a particular experiment, we would use Eq. (13.25), i.e. (13.4), and standard state II, which would make $\mu_i^\ominus$ temperature and pressure dependent:

$$\mu_i = \mu_i^\ominus + RT \ln x_i$$

We would thus obtain

$$\boxed{-\Sigma_i v_i \mu_i^\ominus = RT \ln K_x} \tag{14.23}$$

where $K_x = x_C^c x_D^d / x_A^a x_B^b$ or, more generally, $K_x = \prod_i x_i^{v_i}$. It can be seen from Eq. (14.23) that the equilibrium constant in terms of mole fractions, K_x, must be temperature and pressure dependent, since so is $\mu_i^\ominus$. We may write

$$\boxed{-\Delta G^\ominus = RT \ln K_x} \tag{14.24}$$

RELATIONSHIP BETWEEN THE EQUILIBRIUM CONSTANTS $K_{P/P^\ominus}$ AND K_x FOR AN IDEAL GAS MIXTURE

Consider the equilibrium constant K_x. Employing Dalton's law of partial pressures,

$$x_i = \frac{P_i}{P}$$

Hence,

$$K_x = \frac{(P_C/P)^c (P_D/P)^d}{(P_A/P)^a (P_B/P)^b} \quad \text{or} \quad K_x = \prod_i \left(\frac{P_i}{P}\right)^{v_i}$$

i.e.

$$K_x = \frac{P_C^c P_D^d}{P_A^a P_B^b} \cdot P^{-c-d+a+b} \quad \text{or} \quad K_x = P^{-\Sigma_i v_i} \Pi_i P_i^{v_i}$$

That is,

$$K_x = K_{P/P\ominus} \left(\frac{P}{P^\ominus}\right)^{-\Delta n} \tag{14.25}$$

where $\Delta n = c + d - (a+b)$ or, more generally, $\Delta n = \Sigma_i v_i$.
For the particular case where $\Delta n = 0$, we have $K_{P/P\ominus} = K_x$.

14.5   Homogeneous reactions of non-ideal gases

For a homogeneous gaseous reaction, we allow for the fact that the gases are not ideal by employing an equation for the chemical potential in terms of fugacity, i.e.

$$\mu_i = \mu_i^\ominus + RT \ln \frac{f_i}{f_i^\ominus}$$

and hence obtain

$$-\Sigma_i v_i \mu_i^\ominus = RT \ln K_{f/f\ominus} \tag{14.26}$$

where $K_{f/f\ominus}$ is the equilibrium constant in terms of fugacities. Employing standard state I, $\mu_i^\ominus$ is independent of pressure, which then means that $K_{f/f\ominus}$ is also independent of pressure, from Eq. (14.26).

$$K_{f/f\ominus} = \frac{f_C^c f_D^d}{f_A^a f_B^b}(f^\ominus)^{-\Delta n} = \frac{\phi_C^c \phi_D^d}{\phi_A^a \phi_B^b} \cdot \frac{P_C^c P_D^d}{P_A^a P_B^b}(f^\ominus)^{-\Delta n}$$

or

$$K_{f/f\ominus} = \Pi_i \left(\frac{f_i}{f^\ominus}\right)^{v_i} = \Pi_i \phi_i^{v_i}\left(\frac{P_i}{f^\ominus}\right)^{v_i}$$

where ϕ_i is the fugacity coefficient of i. Hence,

$$K_{f/f\ominus} = K_\phi \cdot K_{P/P\ominus} \tag{14.27}$$

If we were to make the approximation that the gases behave ideally, then the equilibrium constant would be $K_{P/P\ominus}$. However, since the gases are not really ideal, this is not the true equilibrium constant, which is, in fact, $K_{f/f}\ominus$. As mentioned previously, $K_{f/f}\ominus$ is pressure independent, whereas $K_{P/P}\ominus$ must in this case depend on the total pressure, since in Eq. (14.27) K_ϕ is pressure dependent.

14.6 Homogeneous reactions in solution

By substituting the appropriate chemical potentials into Eq. (14.16), we may obtain an expression in the equilibrium constant of our choice. For example, considering an ideal solution, the basic condition for chemical equilibrium is

$$\Sigma_i \, \nu_i \, \mu_i = 0$$

We may use Eq. (13.25) for each component, i.e.

$$\mu_i = \mu_i^\ominus + RT \ln x_i$$

Hence,

$$-\Sigma_i \, \nu_i \, \mu_i^\ominus = RT \ln K_x \qquad (14.28)$$

where $K_x = \Pi_i \, x_i^{\nu_i}$. Alternatively, if we use molality as our measure of composition,

$$\mu_i = \mu_i^{\ominus m} + RT \ln m_i$$

giving

$$-\Sigma_i \, \nu_i \, \mu_i^{\ominus m} = RT \ln K_m \qquad (14.29)$$

where $K_m = \Pi_i \, m_i^{\nu_i}$.

As can be seen from Eqs (14.28) and (14.29), the equilibrium constants K_x and K_m are pressure and temperature dependent. Of course, for non-ideal solutions Eq. (14.8) is true, the standard chemical potentials being based on whatever standard state is suited to our convenience, e.g. standard state III or standard state IV or any other which we might care to decide upon. Eq. (14.28) or Eq. (14.29) could, of course, have been obtained by merely substituting into Eq. (14.8).

14.7 Heterogeneous reactions

We shall first consider heterogeneous reactions involving a gas phase in conjunction with pure liquid phases and/or pure solid phases. By pure liquids we mean that, if more than one liquid is present in the reaction vessel, then these liquids are immiscible, i.e. they do not form a solution; and similarly for solids, no solid solution is formed.

Examples (1) $CaCO_3(s) = CaO(s) + CO_2(g)$
(2) $C(s) + CO_2(g) = 2CO(g)$
(3) $C(s) + H_2O(g) = CO(g) + H_2(g)$

Strictly speaking, we require that the gaseous substances also be immiscible in the condensed phases. Applying Eq. (14.16) or Eq. (14.5) to reaction (1), we have

$$\mu_{CaO} + \mu_{CO_2} - \mu_{CaCO_3} = 0$$

The chemical potential of the gaseous CO_2 may be given by

$$\mu_{CO_2} = \mu_{CO_2}^{\ominus} + RT \ln \frac{P_{CO_2}}{P_{CO_2}^{\ominus}}$$

if the CO_2 is assumed to be an ideal gas. In general, the CO_2 gas need not behave ideally, in which case a similar equation, in which P_{CO_2} is replaced by the fugacity of the CO_2, would be used to give the chemical potential. However, for simplicity, we shall assume the CO_2 gas to be ideal.

Now, at low pressures, we may make the reasonable approximation that the chemical potentials of a condensed substance at 101 325 N m^{-2} and at the pressure P (P less than about 1 MN m^{-2}) at a given temperature are equal (see Eq. 11.32).

$$\mu_{CaO} = \mu_{CaO}^{\ominus}$$

$$\mu_{CaCO_3} = \mu_{CaCO_3}^{\ominus}$$

Hence, we have that, approximately,

$$\mu_{CaO} + \mu_{CO_2} - \mu_{CaCO_3} = \mu_{CaO}^{\ominus} + \mu_{CO_2}^{\ominus} + RT \ln \frac{P_{CO_2}}{P^{\ominus}} - \mu_{CaCO_3}^{\ominus}$$

$$\mu_{CaO} + \mu_{CO_2} - \mu_{CaCO_3} = \mu_{CaO}^{\ominus} + \mu_{CO_2}^{\ominus} - \mu_{CaCO_3}^{\ominus} + RT \ln \frac{P_{CO_2}}{P^{\ominus}}$$

$$\Delta G = \Delta G^{\ominus} + RT \ln \frac{P_{CO_2}}{P^{\ominus}}$$

At equilibrium, $\Delta G = 0$; hence,

$$-\Delta G^{\ominus} = RT \ln \frac{P_{CO_2}}{P^{\ominus}}$$

We therefore have that, at low total pressures, P_{CO_2} is a constant at a given temperature and is independent of pressure since so is $\Delta G^{\ominus}$. P_{CO_2} may be regarded, therefore, as the equilibrium constant at low pressures, $P_{CO_2} \doteqdot K'$, where the prime denotes a pseudo equilibrium constant.

Alternatively, we may consider the chemical potential of a component i to be given, in general, by

$$\mu_i = \mu_i^{\ominus} + RT \ln a_i$$

where the standard state is taken to be pressure independent as is usual for gases and pure substances. Hence,

$$\mu_{CaO} + \mu_{CO_2} - \mu_{CaCO_3} = \mu_{CaO}^\ominus + \mu_{CO_2}^\ominus - \mu_{CaCO_3}^\ominus + RT \ln \frac{a_{CaO} a_{CO_2}}{a_{CaCO_3}}$$

Employing the equilibrium condition $\Delta G = 0$,

$$-(\mu_{CaO}^\ominus + \mu_{CO_2}^\ominus - \mu_{CaCO_3}^\ominus) = RT \ln \left\{ \frac{a_{CaO} a_{CO_2}}{a_{CaCO_3}} \right\}_{eq}$$

i.e.
$$-\Delta G^\ominus = RT \ln K$$

At *low pressures*, the activity of a *pure* solid or liquid is unity (see page 215) and therefore $a_{CaO} = a_{CaCO_3} = 1$. Hence,

$$K = \frac{a_{CaO} a_{CO_2}}{a_{CaCO_3}} \div a_{CO_2} = K'$$

If the CO_2 were an ideal gas, we would have

$$a_{CO_2} = \frac{f_{CO_2}}{f_{CO_2}^\ominus} = \frac{P_{CO_2}}{P_{CO_2}^\ominus}$$

For reaction (2), Eq. (14.6) gives

$$2\mu_{CO} - \mu_C - \mu_{CO_2} = 2\mu_{CO}^\ominus - \mu_C^\ominus - \mu_{CO_2}^\ominus + RT \ln \frac{a_{CO}^2}{a_C a_{CO_2}}$$

or,
$$\Delta G = \Delta G^\ominus + RT \ln \frac{a_{CO}^2}{a_C a_{CO_2}}$$

The equilibrium condition $\Delta G = 0$ gives

$$-\Delta G^\ominus = RT \ln K$$

where
$$K = \frac{a_{CO}^2}{a_C a_{CO_2}} \div \frac{a_{CO}^2}{a_{CO_2}} = K'$$

since $a_C = 1$ at low pressures. Now

$$K' = \frac{a_{CO}^2}{a_{CO_2}} = \left(\frac{f_{CO}}{f_{CO}^\ominus} \right)^2 \cdot \frac{f_{CO_2}^\ominus}{f_{CO_2}} = \frac{f_{CO}^2}{f_{CO_2}} \cdot \frac{f_{CO_2}^\ominus}{(f_{CO}^\ominus)^2}$$

If the standard state I has been chosen, then the fugacities of CO and CO_2 in the standard state are the same. Writing $f_{CO}^\ominus = f_{CO_2}^\ominus = f^\ominus$, we have

$$K' = \frac{f_{CO}^2}{f_{CO}} \cdot \frac{1}{f^\ominus}$$

If the CO and CO_2 were ideal gases, we would have

$$K' = \frac{P_{CO}^2}{P_{CO_2}} \cdot \frac{1}{P^{\ominus}}$$

Consider the reaction

$$C_2H_5OH(l) + 3O_2(g) = 2CO_2(g) + 3H_2O(l)$$

The equilibrium condition $\Delta G = 0$ gives, for this reaction,

$$2\mu_{CO_2} + 3\mu_{H_2O} - \mu_{C_2H_5OH} - 3\mu_{O_2} = 0 \qquad (14.30)$$

To derive the equilibrium constant for the reaction, we must decide which are the most convenient units of measurement for each substance, so that a convenient standard state for each substance may be used. For instance, for the gases – that is, the oxygen and carbon dioxide – either partial pressure (or fugacity) may be used, so that standard state I is a convenient one:

$$\mu_{O_2} = \mu_{O_2}^{\ominus} + RT \ln \frac{f_{O_2}}{f^{\ominus}}$$

$$\mu_{CO_2} = \mu_{CO_2}^{\ominus} + RT \ln \frac{f_{CO_2}}{f^{\ominus}}$$

For water, mole fraction is the usual measure of composition; hence, standard state III could be used:

$$\mu_{H_2O} = \mu_{H_2O}^{\ominus} + RT \ln a_{x_{H_2O}}$$

and for ethanol we could decide to use molality:

$$\mu_{C_2H_5OH} = \mu_{C_2H_5OH}^{\ominus} + RT \ln a_{m_{C_2H_5OH}}$$

Substituting these expressions for the chemical potentials in Eq. (14.30),

$$-\Delta G^{\ominus} = RT \ln K$$

where

$$\Delta G^{\ominus} = 2\mu_{CO_2}^{\ominus} + 3\mu_{H_2O}^{\ominus} - \mu_{C_2H_5OH}^{\ominus} - 3\mu_{O_2}^{\ominus}$$

and the equilibrium constant is given by

$$K = \frac{f_{CO_2}^2 a_{x_{H_2O}}^3 f^{\ominus}}{a_{m_{C_2H_5OH}} f_{O_2}^3}$$

This equilibrium constant is both temperature and pressure dependent.

If we make the assumption that the pressure is not large, so that the gases behave ideally, and that the solution is dilute (i.e. $x_{H_2O} \doteqdot 1$) and, hence, ideal, then

$$K \doteqdot \frac{P_{CO_2}^2 \cdot P^\ominus}{m_{C_2H_5OH}P_{O_2}^3}$$

14.8 Temperature dependence of the equilibrium constant

Rearranging and differentiating the van't Hoff isotherm, we have

$$\left(\frac{\partial \ln K}{\partial T}\right)_P = -\left(\frac{\partial \dfrac{\Delta G^\ominus}{RT}}{\partial T}\right)_P$$

Employing the Gibbs–Helmholtz equation (Eq. 8.23), we have:

$$\boxed{\left(\frac{\partial \ln K}{\partial T}\right)_P = \frac{\Delta H^\ominus}{RT^2}} \qquad (14.31)$$

$\Delta H^\ominus$ is the standard heat of reaction at the temperature T and is the enthalpy change for the reaction when the reactants and products are in their standard states, i.e.

$$\Delta H^\ominus = cH_C^\ominus + dH_D^\ominus - aH_A^\ominus - bH_B^\ominus \quad \text{or} \quad \Delta H^\ominus = \Sigma_i v_i H_i^\ominus$$

If a reactant or product is a pure substance, then from the usual standard state, $H_i^\ominus = h_i$, where h_i is the molar enthalpy of i. If a reactant or product is a component of a solution, then obviously

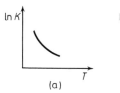

Figure 14.3.
(a) Exothermic reaction;
(b) endothermic reaction

the standard state is defined as that for a component of a solution, i.e. standard states III and IV, and $H_i^\ominus = \bar{H}_i^\ominus$.

Equation (14.31) is referred to as the van't Hoff equation. It is important to remember that if a reaction is exothermic $\Delta H^\ominus$ is negative (see Chapter 2).

Integrating Eq. (14.31) from temperature T_1 to T_2 at a given pressure,

$$\int_{K_1}^{K_2} d \ln K = \int_{T_1}^{T_2} \frac{\Delta H^\ominus}{RT^2} dT$$

If, over *small* temperature ranges, $\Delta H^{\ominus}$ may be assumed approximately constant, we have

$$\ln \frac{K_2}{K_1} = -\frac{\Delta H^{\ominus}}{R}\left(\frac{1}{T_2}-\frac{1}{T_1}\right)$$

(14.32)

For *larger* temperature changes, i.e. when $\Delta H^{\ominus}$ cannot be assumed constant, use is made of the Kirchhoff equation (see Eq. 2.73) to determine the temperature dependence of ΔH:

$$\left(\frac{\partial \Delta H}{\partial T}\right)_P = \Delta C_P$$

If the molar heat capacity of each component i is expressed as a power series (this being the usual form for expressing the experimental results for heat capacities), we have

$$c_P^{\ominus} = \alpha + \beta T + \gamma T^2 + \dots$$

(14.33)

Now
$$\Delta C_P = c(c_P)_C + d(c_P)_D - a(c_P)_A - b(c_P)_B$$

That is,

$$\Delta C_P = \Sigma_i v_i(\alpha_i + \beta_i T + \gamma_i T^2 + \dots)$$

(14.34)

The Kirchhoff equation may therefore be written as

$$\left(\frac{\partial \Delta H}{\partial T}\right)_P = \Sigma_i v_i(\alpha_i + \beta_i T + \gamma_i T^2 + \dots)$$

Integrating at a given pressure,

$$\Delta H = \Delta H_0 + \Sigma_i v_i\left(\alpha_i T + \frac{\beta_i T^2}{2} + \frac{\gamma_i T^3}{3} + \dots\right)$$

(14.35)

where ΔH_0 is the constant of integration and may be evaluated if the heat of reaction is known at a given temperature.

Using this equation to determine the standard enthalpy change, and substituting for ΔH in the van't Hoff equation, Eq. (14.31),

$$R\left(\frac{\partial \ln K}{\partial T}\right)_P = \frac{\Delta H_0}{T^2} + \Sigma_i v_i\left(\frac{\alpha_i}{T} + \frac{\beta_i}{2} + \frac{\gamma_i}{3}T + \dots\right)$$

Integrating at the given pressure,

$$R \ln K = \mathscr{C} - \frac{\Delta H_0}{T} + \Sigma_i v_i\left(\alpha_i \ln T + \frac{\beta_i}{2}T + \frac{\gamma_i}{6}T^2 + \dots\right)$$

(14.36)

where $\mathscr{C}$ is the constant of this integration, which may be determined if ΔH_0 and K are known at a given temperature. This equation may be used to determine the equilibrium constant at a given temperature. By also employing the van't Hoff isotherm (Eq. 14.8), the change in the standard Gibbs function for the reaction may be found at a given temperature.

14.9 Pressure dependence of the equilibrium constant

The equilibrium constant of a reaction may or may not be pressure dependent according to the definitions of the standard states of the species involved. The standard states may be defined so as to be independent of the pressure of the system, so that $\Delta G^{\ominus}$, and thus K, will also be pressure independent. If the standard state of all species is defined as the state of the pure substance at a pressure of 101 325 N m^{-2} at the temperature under consideration, or more strictly as being at a fugacity of 101 325 N m^{-2}, then the equilibrium constant is pressure independent.

If, however, the definition of the standard state of one specie is the state of the pure substance at the *pressure P* at the temperature under consideration, then the equilibrium constant is dependent on pressure.

Differentiating the van't Hoff equation with respect to pressure at constant temperature,

$$-\left(\frac{\partial \Delta G^{\ominus}}{\partial P}\right)_T = RT\left(\frac{\partial \ln K}{\partial P}\right)_T$$

Employing Eq. (8.17), we have

$$\left(\frac{\partial \ln K}{\partial P}\right)_T = -\frac{\Delta V^{\ominus}}{RT} \qquad (14.37)$$

where $\Delta V^{\ominus}$ is the difference in volume of the products and reactants in their standard states and is the standard volume change.

14.10 Le Chatelier's principle

If one of the variables affecting the position of equilibrium of a system is altered, then the position of the equilibrium is displaced in such a direction as to oppose the change.

Le Chatelier's principle is probably more familiar to us as the

Law of Cussedness! For example, if we try to push a box along a horizontal surface, friction tends to oppose the motion of the box. Lenz's law in electricity is another example. Let us now examine the chemical implications of the principle and also see if our thermo-dynamically derived formulae giving the temperature and pressure dependence of the equilibrium constant are in agreement with it.

For an exothermic reaction, $\Delta H^\ominus$ is a negative quantity (see Section 2.19). Hence, from Eq. (14.31), $(\partial \ln K / \partial T)_P$ is negative – that is, with increase in temperature the equilibrium constant decreases. This is in agreement with Le Chatelier's principle, since at a higher temperature an exothermic reaction would not favour the formation of products and would thus oppose the rise in temperature.

If, for a reaction at a given temperature and pressure, the formation of products from reactants leads to a decrease in volume, $\Delta V^\ominus$ is a negative quantity, and hence in Eq. (14.37) $(\partial \ln K / \partial P)_T$ is positive. That is, for an increase in pressure the equilibrium constant also increases. According to Le Chatelier's principle, increase in pressure would result in the formation of product, since this would oppose the pressure increase. For example, for the reaction

$$N_2 + 3H_2 = 2NH_3$$

the volume of the ammonia is less than that of the reactants. Hence, to produce ammonia, high pressures are employed. This reaction is endothermic; thus increase in temperature results in a decrease in the yield of ammonia. Hence, in the industrial production of ammonia, it is desirable to have a low temperature. On the other hand, if the temperature is too low, a reasonable rate of reaction is not achieved, so that usual operating conditions are around 770 K and 20 000 000 Nm^{-2} (200 atm).

14.11   Methods for the determination of changes in the Gibbs function

(1) By the addition and subtraction of available values of the standard Gibbs functions of formation, e.g.

$$
\begin{array}{ll}
C + \tfrac{1}{2}O_2 = CO & \Delta G_1^\ominus \\
\underline{CO + \tfrac{1}{2}O_2 = CO_2} & \underline{\Delta G_2^\ominus} \\
C + O_2 = CO_2 & \Delta G_3^\ominus = \Delta G_1^\ominus + \Delta G_2^\ominus
\end{array}
$$

(2) From the equilibrium constant, by employing the van't Hoff equation, $\Delta G^\ominus = -RT \ln K$. This is only possible if (a) side reactions do not occur; (b) the equilibrium position is not too close

to one end of the composition range for the reaction; (c) the reaction is reasonably fast or can be made so by the addition of catalyst.

(3) From e.m.f. measurements (see Section 15.9).

(4) From the equation $\Delta G^\ominus = \Delta H^\ominus - T\Delta S^\ominus$, using values of $\Delta S^\ominus$, which may be obtained by means either of the third law and calorimetric techniques (see Section 7.3) or of statistical methods (see Section 16.11), and $\Delta H^\ominus$, which may usually be easily determined calorimetrically (see Section 2.21) or by using Hess' law (see Section 2.20) or by e.m.f. measurements employing the Gibbs–Helmholtz equation (see Section 15.9).

Example 1 The equilibrium constant $K_{P/P^\ominus}$ for the reaction $CO_2 = CO + \frac{1}{2}O_2$ is 0·155 at 3000 K. Calculate the partial pressures and composition of the substituents at equilibrium at 1·013 25 MN m^{-2}.

Let the fraction of 1 mole of CO_2 which dissociates be x, so that at equilibrium $1-x$ moles CO_2 remain and x moles CO and $x/2$ moles O_2 have been formed.

The total amount of substance in moles is $1-x+x+x/2 = 1+x/2$.
From Dalton's law, the partial pressures are

$$P_{CO_2} = \frac{1-x}{1+x/2} \times 1\,013\,250 \text{ N m}^{-2}$$

$$P_{CO} = \frac{x}{1+x/2} \times 1\,013\,250 \text{ N m}^{-2}$$

$$P_{O_2} = \frac{x/2}{1+x/2} \times 1\,013\,250 \text{ N m}^{-2}$$

Hence,

$$K_{P/P^\ominus} = \frac{\dfrac{\dfrac{x}{1+x/2} \cdot 1\,013\,250}{P^\ominus} \cdot \left(\dfrac{\dfrac{x/2}{1+x/2} \cdot 1\,013\,250}{P^\ominus}\right)^{\frac{1}{2}}}{\dfrac{\dfrac{1-x}{1+x/2} \cdot 1\,013\,250}{P^\ominus}}$$

$P^\ominus = 101\,325$ N m^{-2} so that

$$0{\cdot}155 = \frac{x^{\frac{3}{2}}\,10^{\frac{1}{2}}}{(2+x)^{\frac{1}{2}}(1-x)}$$

giving $x = 0{\cdot}16$. Using this value of x, we may obtain the equilibrium composition and the partial pressures.

Example 2   Calculate the enthalpy change, the standard Gibbs function change and the equilibrium constant at 723 K for the reaction

$$\tfrac{1}{2}N_2(g) + \tfrac{3}{2}H_2(g) = NH_3(g) \qquad \Delta H^\circ(293\ K) = -45\cdot751\ kJ$$

given that in the temperature range considered

$$c_P(NH_3(g)) = 28\cdot03 + 0\cdot02635T\ J\ mol^{-1}$$
$$c_P(N_2(g)) = 27\cdot17 + 0\cdot00418T\ J\ mol^{-1}$$
$$c_P(H_2(g)) = 27\cdot67 + 0\cdot00339T\ J\ mol^{-1}$$

and that the standard Gibbs function change at 298 K is $-16\cdot180$ kJ. Employing the heat capacities given, it is found that

$$\Delta c_P = -27\cdot10 + 0\cdot0192T$$

Substituting this into the Kirchhoff equation, Eq. (2.73), and integrating from 293 K to temperature T, we have

$$\Delta H^\circ - \Delta H^\circ(293\ K) = \int_{293}^{T} \Delta c_P\, dT$$

$$\Delta H^\circ = -38690 - 27\cdot10T + 0\cdot0096T^2\ J$$

Hence, $\Delta H^\circ(723\ K) = -53\cdot231$ kJ. From Eq. (8.23),

$$\int_{T=298}^{723} d\left(\frac{\Delta G^\circ}{T}\right) = -\int_{298}^{723} \frac{\Delta H^\circ}{T^2}\, dT$$

$$\frac{\Delta G^\circ}{723} - \frac{-16180}{298} = -\int_{298}^{723}\left(\frac{-38690}{T^2} - \frac{27\cdot10}{T} + 0\cdot0096\right) dT$$

Hence, $\Delta G^\circ = 29\cdot60$ kJ mol^{-1}. Substituting this value into the van't Hoff equation, i.e.

$$-\Delta G^\circ = RT \ln K_{P/P^\circ}$$

we have

$$-29600 = 8\cdot314 \times 723 \times \ln K_{P/P^\circ}$$

giving

$$K_{P/P^\circ} = 0\cdot00728$$

Problems

$$R = 8\cdot314\ J\ K^{-1}\ mol^{-1}$$

1.   The equilibrium constant K_{P/P° for the reaction $CO + \tfrac{1}{2}O_2 = CO_2$

at 2000 K is 715 and at 3000 K, 3·056. Calculate the mean standard enthalpy change for the reaction.
Answer $-1·23 \times 10^8$ J

2. The equilibrium constant $K_{P/P^\ominus}$ for the reaction $N_2 + 3H_2 = 2NH_3$ at 770 K is $1·44 \times 10^{-3}$. The standard enthalpy change, $\Delta H^\ominus(770 \text{ K})$, is $-105·1$ kJ. Calculate the equilibrium constant at 673 K.
Answer $1·65 \times 10^{-4}$

3. For the gaseous reaction $H_2 + \frac{1}{2}O_2 = H_2O$, the standard Gibbs function change is -137 kJ at 2000 K. Calculate the equilibrium constant at this temperature.
Answer $1·51 \times 10^7$

4. Determine the composition and partial pressures at equilibrium of the components of the reaction $CO_2 = CO + \frac{1}{2}O_2$.
$K_{P/P^\ominus}$ (3000 K) $= 0·155$ at a pressure of 101 325 N m^{-2}

Answer $x_{CO_2} = 0·610$, $x_{CO} = 0·260$, $x_{CO_2} = 0·130$
$P_{CO_2} = 61\ 800$ N m^{-2}, $P_{CO} = 26\ 350$ N m^{-2}
$P_{O_2} = 13\ 160$ N m^{-2}

5. The enthalpy change accompanying the phase change of monoclinic sulphur to rhombic sulphur at 368·7 K is -397 J mol^{-1} and the molar heat capacities in the temperature range considered are

$$c_P(\text{rhombic}) = 14·60 + 0·02658\,T \text{ J K}^{-1} \text{ mol}^{-1}$$
$$c_P(\text{monoclinic}) = 14·88 + 0·02919\,T \text{ J K}^{-1} \text{ mol}^{-1}$$

Show that $\Delta H^\ominus = -127 - 0·263\,T - 0·001296\,T^2$ J mol^{-1} and that the change in the standard Gibbs function accompanying the transformation at 298 K is -67 J mol^{-1}. (Remember that $\Delta G^\ominus$ (368·7 K) $= 0$).

6. The equilibrium constant $K_{P/P^\ominus}$ for the water gas reaction

$$CO(g) + H_2O(g) = H_2(g) + CO_2(g)$$

is $9·53 \times 10^4$ at 298 K. Given that the standard Gibbs functions of $CO_2(g)$ and $CO(g)$ are $-394\ 000$ J mol^{-1} and $-137\ 000$ J mol^{-1} at 298 K, calculate the standard Gibbs function of water.
Answer $-228\ 000$ J mol^{-1}

15
Electrolytes

15.1 Introduction

In solution, electrolytes conduct electricity due to their dissociating into charged particles, known as ions, which move in an applied electric field. The positive ions are known as cations and the negative ions as anions because in the applied field they seek the cathode and anode, respectively. The solution as a whole is, of course, electrically neutral. Strong acids, strong bases and their salts are strong electrolytes since they are completely dissociated in solvents such as water and liquid ammonia. Weak electrolytes such as acetic acid and ammonia are not completely dissociated in solution. The undissociated molecules exist in equilibrium with the ions.

Consider the dissociation of an electrolyte,

$$M_{v_+} A_{v_-} = v_+ M^{Z+} + v_- A^{Z-} \tag{15.1}$$

where v_+ and v_- are the numbers of positive and negative ions, respectively, which are obtained when one molecule of electrolyte dissociates, and Z_+ and Z_- are the charges of the ions. Since the parent electrolyte was not charged, eletrical neutrality dictates that

$$v_+ Z_+ + v_- Z_- = 0 \tag{15.2}$$

For example, for

$$H_2SO_4 = 2H^+ + SO_4^{--}$$

we have

$$v_+ = 2, v_- = 1, Z_+ = 1, Z_- = -2$$

and for

$$MgCl_2 = Mg^{++} + 2Cl^-$$

we have

$$v_+ = 1, v_- = 2, Z_+ = 2, Z_- = -1$$

Consider a solution prepared by dissolving n moles of electrolyte $M_{v_+} A_{v_-}$ in n_s moles of solvent. Then,

$$n_+ = v_+(n-n_u) \qquad (15.3)$$
$$n_- = v_-(n-n_u) \qquad (15.4)$$

where n_+, n_- and n_u are the number of moles of cation, anion and undissociated molecule in the solution. For a strong electrolyte, $n_u = 0$, e.g. 1 mole of hydrochloric acid contains 1 mole of chloride ion, 1 mole of sulphuric acid contains 2 moles of hydrogen ion.

The discussion which follows will be quite general and apply to both weak and strong electrolytes.

15.2   Chemical potential

For mathematical convenience, it is usual to define chemical potentials for the different ions in solution, e.g. the chemical potential of the cation is

$$\mu_+ = \left(\frac{\partial G}{\partial n_+}\right)_{T,P,n_-,n_s} \qquad (15.5)$$

where n_s is the number of moles of solvent. These ionic chemical potentials, although useful, lack experimental reality as it is impossible to vary n_+ without altering n_- since electrical neutrality is always maintained.

For an infinitesimal change in the Gibbs function of the solution at constant pressure and temperature, we have

$$dG = \mu_u \, dn_u + \mu_+ \, dn_+ + \mu_- \, dn_- + \mu_s \, dn_s \qquad (15.6)$$

Employing Eqs (15.3) and (15.4), this becomes

$$dG = \mu_u \, dn_u + v_+\mu_+(dn - dn_u) + v_-\mu_-(dn - dn_u) + \mu_s \, dn_s \qquad (15.7)$$

or $dG = (\mu_u - v_+\mu_+ - v_-\mu_-)\, dn_u + (v_+\mu_+ + v_-\mu_-)\, dn + \mu_s \, dn_s$

Now, for a closed system, $dn = dn_s = 0$. Also, when the dissociation equilibrium is reached, $dG = 0$. Hence, since dn_u is an infinitesimal arbitrary change, we have

$$\mu_u = v_+\mu_+ + v_-\mu_- \qquad (15.8)$$

Now consider an open system to which electrolyte and solvent are added so slowly that dissociation equilibrium is maintained at all times. Equation (15.7) then gives

$$dG = (v_+\mu_+ + v_-\mu_-)\,dn + \mu_s\,dn_s \qquad (15.9)$$

The chemical potential of the electrolyte is defined as

$$\mu = \left(\frac{\partial G}{\partial n}\right)_{T,P,n_s} \qquad (15.10)$$

It is therefore seen, on comparing Eq. (15.10) with Eq. (15.9), that

$$\boxed{\mu = v_+\mu_+ + v_-\mu_-} \qquad (15.11)$$

Although μ_+ and μ_- lack experimental significance, μ does not, since there are no restrictions on the variation of n.

15.3 Activity

Electrolyte concentrations have almost always in the past been measured in terms of molality. The electrolyte solution is considered to approach ideal behaviour at infinite dilution.

The chemical potential of the cation is given by

$$\mu_+ = \mu_+^\ominus + RT \ln a_+ \qquad (15.12)$$

The activity of the cation a_+ may be written in terms of the activity coefficient, γ_+, and the molality of the cation, m_+;

$$\mu_+ = \mu_+^\ominus + RT \ln \gamma_+ m_+ \qquad (15.13)$$

$\gamma_+ \rightarrow 1$ as $m_+ \rightarrow 0$, i.e. as $m \rightarrow 0$.
The chemical potential of the solution is found by employing Eq. (15.11),

$$\mu = \mu^\ominus + v_+ RT \ln \gamma_+ m_+ + v_- RT \ln \gamma_- m_-$$

where

$$\mu^\ominus = v_+\mu_+^\ominus + v_-\mu_-^\ominus.$$

This equation may be rewritten as

$$\boxed{\mu = \mu^\ominus + vRT \ln \gamma_\pm m_\pm} \qquad (15.14)$$

where

$$v = v_+ + v_- \tag{15.15}$$

$$\gamma_\pm^v = \gamma_+^{v_+} \gamma_-^{v_-} \tag{15.16}$$

$$m_\pm^v = m_+^{v_+} m_-^{v_-} \tag{15.17}$$

$\gamma_\pm$ is known as the *mean activity coefficient* and $m_\pm$ as the *mean ionic molality*. The *mean ionic activity* $a_\pm$ is given by

$$\boxed{a_\pm = \gamma_\pm m_\pm} \tag{15.18}$$

For a 1:1 electrolyte such as NaCl, $\gamma_\pm = (\gamma_+ \gamma_-)^{\frac{1}{2}}$ and $m_\pm = (m_+ m_-)^{\frac{1}{2}}$. Therefore, for a strong 1:1 electrolyte, $m_\pm$ is equal to m, the molality of the electrolyte solution. For a 1:2 electrolyte, such as $CaCl_2$, $\gamma_\pm = (\gamma_+ \gamma_-^2)^{\frac{1}{3}}$ and $m_\pm = (m_+ m_-^2)^{\frac{1}{3}}$. Therefore, for a strong 1:2 electrolyte, $m_\pm = 2^{\frac{2}{3}} m$.

In a mixture of strong electrolytes with a common ion, such as NaCl and Na_2SO_4, the mean ionic molalities are

$$m_{\pm NaCl} = [m_{Na^+} . m_{Cl^-}]^{\frac{1}{2}} = [(m_{NaCl} + 2m_{Na_2SO_4})m_{NaCl}]^{\frac{1}{2}}$$

and

$$m_{\pm Na_2SO_4} = [m_{Na^+}^2 . m_{SO_4^{2-}}]^{\frac{1}{3}} = [(m_{NaCl} + 2m_{Na_2SO_4})2m_{Na_2SO_4}]^{\frac{1}{3}}$$

where m_{NaCl} and $m_{Na_2SO_4}$ are the molalities of the NaCl and Na_2SO_4 solutions.

For a strong electrolyte, $\gamma_\pm$ is a measurable property, even though γ_+ and γ_- are not, since in Eq. (15.14) all the other properties are measurable.

15.4 Solubility product

Consider the equilibrium of a saturated electrolyte and its undissociated solid electrolyte at a given temperature and pressure. At equilibrium the chemical potential of the electrolyte in solution must equal the chemical potential of the solid undissociated electrolyte,

$$\mu_u' = \mu = \mu^\ominus + v \, RT \ln \gamma_\pm m_\pm \tag{15.19}$$

Since μ_u' and $\mu^\ominus$ for the electrolyte are constant at a given T and P, we have

$$\boxed{(\gamma_\pm m_\pm)^v = \text{S.P.}} \tag{15.20}$$

where S.P. is a constant known as the solubility product of the electrolyte. For an ideal solution, $\gamma_{\pm} = 1$ and Eq. (15.20) becomes

$$m_+^{v_+} m_-^{v_-} = \text{S.P.} \qquad (15.21)$$

This equation gives an explanation of the *common ion effect*, in which an electrolyte is precipitated from its saturated solution by the addition of another electrolyte having a common ion. For example, NaCl is precipitated from its saturated solution by the addition of HCl. This occurs because the concentration of the common ion has increased so that the solubility product is exceeded.

15.5   Deviations from ideal behaviour

Deviations from ideal behaviour are due to the interaction forces between the molecules not being uniform throughout the solution, i.e. the forces between solvent–solvent, solvent–solute and solute–solute being dissimilar. In a non-electrolyte solute, the forces between the uncharged molecules (known as van der Waals' forces) vary as the inverse of the seventh power of the separation of the molecules. Therefore the forces act only over a short range, so that in a dilute solution, where the solute molecules are surrounded by solvent molecules, the forces are uniform and the solution behaves ideally. However, in electrolyte solutions the forces between the ions are coulombic and proportional to $Z_+ Z_- / r^2$, where r is the ion separation. These forces extend over a great distance, so that, unless the solution is *extremely* dilute, such that the ions are widely separated, deviations from ideal behaviour exist. Electrolyte solutions may, in fact, be expected to show very much greater deviations from ideal behaviour than non-electrolyte solutions. Also, the larger the charge of the ions, the greater the deviations become.

The activity coefficient is a measure of the deviation from ideal behaviour and is plotted against molality in Figure 15.1 and against the square root of molality in Figure 15.2. It is seen that the non-electrolyte sucrose remains ideal for dilute solutions. The activity coefficient of uncharged components may be taken as unity in dilute solutions. Also, it is found that the activity coefficients of electrolytes pass through a minimum and then, with increase in concentration, increase to values greater than unity.

Since, in Figure 15.2, the slopes are finite and non-zero for very dilute solutions, this indicates that $\gamma_{\pm}$ depends on $m^{\frac{1}{2}}$.

Lewis in 1913 showed empirically that, for dilute solutions, $\gamma_{\pm}$

was independent of the chemical nature of the ions in solution and that, for all the ions in the solution,

$$\gamma_\pm = 1 - AI^{\frac{1}{2}} \tag{15.22}$$

where A is a constant and

$$I = \tfrac{1}{2}\Sigma_i \, Z_i^2 m_i, \tag{15.23}$$

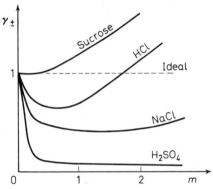

Figure 15.1. Graph of mean ionic activity coefficient against molality

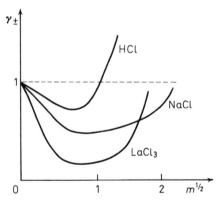

Figure 15.2. Graph of mean ionic activity coefficient against the square root of the molality

Z_i and m_i being, respectively, the number of charges on and the molality of the ion i. I is known as the *ionic strength* and measures the total ionic concentration of the solution, including the effect of ionic charge.

Debye and Hückel in 1923 justified this equation theoretically by assuming that strong electrolytes were completely dissociated in solution and that the observed deviations from ideal behaviour were due to electrical interactions between the ions. The Debye–Hückel limiting law states:

$$\log_{10} \gamma_\pm = -|Z_+ Z_-| A I^{\frac{1}{2}}$$

where A is a constant dependent on the solvent and temperature,

$$A = \left(\frac{e^6 \pi L \rho}{500 \varepsilon^3 k^3 T^3} \right)^{\frac{1}{2}} \tag{15.24}$$

where e is the electronic charge, L is Avogadro's constant, ρ is the solvent density and ε is the solvent dielectric constant. For water at 298 K, $A = 0.509$.

However, only for very dilute solutions ($I = 10^{-3}$) does the limiting law agree with experimental results.

15.6 Dissociation constant

Electrolyte dissociations are merely reactions involving ions and should be treated thermodynamically just as any other reaction. Consider the dissociation of acetic acid,

$$HAc \rightleftharpoons Ac^- + H^+ \quad (Ac = CH_3COO)$$

At equilibrium,

$$\mu_{HAc} = \mu_{Ac^-} + \mu_{H^+}$$

The equilibrium constant is

$$K = \frac{m_{H^+} \cdot m_{Ac^-} \cdot \gamma_\pm^2}{m_{HAc} \cdot \gamma_{HAc}} \tag{15.25}$$

K is known as the dissociation constant.

The activity coefficient of undissociated acetic acid, γ_{HAc}, is unity for dilute solutions. If the solution is very dilute, about 10^{-4} molality, then it may be considered as ideal: $\gamma_\pm = 1$.

Let the molality of the acid be m and the molality of the undissociated acid be $(1-\alpha)m$; then the molality of the H^+ ions is αm and that of the Ac^- ions is αm, where α is the degree of dissociation:

$$K = \frac{m_{H^+} \cdot m_{Ac^-}}{m_{HAc}} = \frac{\alpha^2 m}{1-\alpha}$$

This is known as Ostwald's dilution law.

15.7 Electrochemical cells

Electrochemical cells convert energy liberated in a chemical or physical change into electrical energy. A typical cell is the Daniell cell

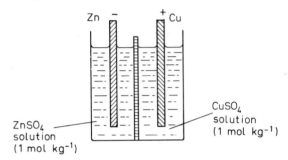

Figure 15.3. Daniell cell

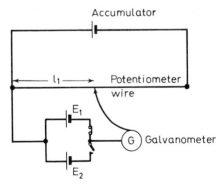

Figure 15.4. Potentiometer

(Figure 15.3), in which a porous pot separates the two solutions shown but allows electrical contact.

Such a cell may be shown diagrammatically as

$$Zn \mid ZnSO_4(1 \text{ M}) \mid CuSO_4(1 \text{ M}) \mid Cu$$

the vertical lines denoting phase boundaries.

When the copper and zinc electrodes are connected, a current flows due to the energy released by the reaction

$$Zn + Cu^{++} \longrightarrow Zn^{++} + Cu$$

The potential across the electrodes measured when the circuit is open, i.e. no current is flowing, is known as the *electromotive force, e.m.f.* The e.m.f. of a cell may be measured by means of a *potentiometer* (Figure 15.4), the accumulator providing a steady current along the potentiometer wire.

The Weston cell (Figure 15.5) is often used as a standard for comparison:

Cd amalgam (12·5%) | 3CdSo$_4$. 8H$_2$O(s) | CdSO$_4$(sat. soln.) | Hg$_2$SO$_4$(s) | Hg

where (s) represents solid. The e.m.f. of the Weston cell at 298 K is 1·01463 V and it decreases by $4·05 \times 10^{-5}$ V per degree rise in temperature. The reaction in the cell is

$$Cd(s) + Hg_2SO_4(s) + \frac{8}{3}H_2O = CdSO_4 . \frac{8}{3}H_2O(s) + 2Hg(l)$$

There are two types of cell. *Cells with liquid junctions*, as in the case of the Daniell cell, are known as *cells with transport*, since electrolyte may be transferred across the boundary. *Cells without liquid junctions*, such as the Weston cell where there is only one electrolyte solution, are known as *cells without transport*.

15.8   Reversible cells

An electrochemical cell is reversible only if (1) when no current passes through the cell, it is in stable equilibrium; (2) when the direction of an infinitesimal current passing through the cell is reversed, all the processes occurring in the cell are also exactly reversed. The potential difference across the electrodes only differs infinitesimally in the three cases (a) when an infinitesimal current passes through the cell, (b) when the direction of this current is reversed, and (c) when no current flows.

For such a reversible cell, any chemical reactions and physical changes occurring in the cell can proceed in either direction. For example, the simple cell,

$$Zn | H_2SO_4 | Cu$$

is not reversible. The reactions

$$Zn = Zn^{++} + 2e \quad \text{and} \quad 2H^+ + 2e = H_2$$

take place spontaneously, even when no current is taken from the cell. Hence, a state of equilibrium is not achieved. In irreversible cells, the original conditions cannot be restored by reversing the current.

A cell can behave reversibly *only* if an infinitesimal current passes through the cell. If large currents pass through it, heat is generated

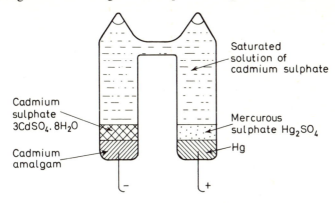

Figure 15.5. Standard Weston cell

and concentration gradients which cannot be obliterated by diffusion, since this is relatively slow, are set up around the electrodes.

15.9 Determination of thermodynamic functions using reversible cells

If an infinitesimal current passes through a reversible cell, the process occurring is thermodynamically reversible and the *work done by* the cell at constant temperature and pressure, which excludes work done by expansion, is equal to the *decrease* in the Gibbs function.

If the e.m.f. of the cell is E volts, the reversible electrical work done when nF coulombs of electricity pass through the cell (where F is the Faraday constant and equals Le, where e is the charge of an electron and L is the Avogadro constant) is nFE joules. Hence,

$$-\Delta G = nFE \qquad (15.26)$$

This has been derived previously in Section 1.14.

By measuring the e.m.f. of a cell, the Gibbs function change for the reaction occurring under the conditions in the cell may be determined, the Gibbs function change thus calculated being for the reactants and products in their particular states in the cell.

We have shown previously (Eq. 8.21) that

$$\Delta S = -\left(\frac{\partial \Delta G}{\partial T}\right)_P$$

Hence,

$$\Delta S = nF\left(\frac{\partial E}{\partial T}\right)_P \qquad (15.27)$$

That is, if the temperature dependence of the e.m.f. of the cell is known, the entropy change for the reaction may be found. From the Gibbs–Helmholtz equation,

$$\Delta H = \Delta G - T\left(\frac{\partial \Delta G}{\partial T}\right)_P$$

Hence,

$$\Delta H = -nF\left[E - T\left(\frac{\partial E}{\partial T}\right)_P\right] \qquad (15.28)$$

Since the enthalpy change may also be found from E and its temperature coefficient, we see that, by measuring the e.m.f. of the cell at various temperatures (at constant P) and plotting E against T, the gradient $(\partial E/\partial T)_P$ may be found at the required temperature.

15.10   Liquid junctions and salt bridges

In all cells with liquid junctions, such as the Daniell cell, a liquid-junction potential occurs. It is found that by using a *salt bridge*, a connecting tube filled with a saturated salt solution such as KCl or NH_4NO_3, this effect is reduced. The liquid-junction potential cannot actually be measured, although calculations involving ionic mobilities show that in favourable cases it is negligible, being only a few millivolts. Salts such as KCl and NH_4NO_3 are chosen for the salt bridge, since the ionic mobilities of their cations and anions are almost identical. By employing a salt bridge, diffusion at the liquid boundary is also prevented.

The salt-bridge solution may be made up as a gel, which not only makes the mechanical handling of the bridge simpler but also prevents diffusion and contamination of the bridge by the electrolyte solutions of the cell. The cell shown in Figure 15.6 may be written

$$Zn \,|\, ZnSO_4 \,\|\, CuSO_4 \,|\, Cu$$

the double lines indicating a salt bridge. However, it is best, if

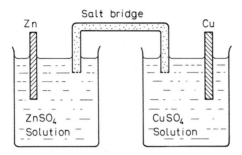

Figure 15.6. Salt bridge

possible, to avoid liquid junctions completely by employing cells which involve a single electrolyte, such as the Weston cell.

15.11 Half cells

The electrodes constituting a reversible cell, together with their appropriate solutions, are known as *half cells*, e.g. $Zn \,|\, ZnSO_4$ (1 M). The combination of *any* two half cells gives a reversible cell. There are three main types of half cell.

Type I – a metal or non-metal in contact with a solution containing its ions.

Examples of metals are:

$$Zn \,|\, Zn^{++}$$
$$Cu \,|\, Cu^{++}$$
$$Ag \,|\, Ag^{+}$$

Sometimes amalgams are used because a metal is not sufficiently pure, e.g. $Na \, amalgam \,|\, Na^{+}$. The advantages gained by using amalgams are that the concentration of the metal may be varied and strains in the metal and polarisation are avoided.

The general reaction that occurs at these electrodes is

$$M \rightleftharpoons M^{+} + e$$

Examples of non-metals are gas electrodes, e.g. hydrogen gas in contact with hydrogen ions, chlorine gas in contact with chloride ions. Because non-metals are poor conductors, an inert metal, such as platinum, is used to provide good electrical contact between the non-metal and solution. For example, in the hydrogen electrode, Pt, $H_2 | H^+$, hydrogen is bubbled over a platinum strip which is immersed in an acid solution (Figure 15.7). The hydrogen probably dissociates into atoms on the platinum and coats its surface. The electrode reaction is

$$\tfrac{1}{2}H_2 \longrightarrow H^+ + e$$

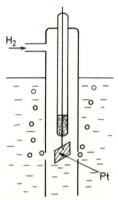

Figure 15.7. Hydrogen electrode

Type II – metal and a sparingly soluble salt of the metal in contact with solution containing the same anion as the salt, e.g.

$$Hg | Hg_2Cl_2(s) | Cl^- \quad \text{(known as calomel electrode)}$$

$$Cd\ amalgam | 3CdSO_4 . 8H_2O | SO_4^{--}$$

$$Ag | AgCl(s) | Cl^-$$

The reaction occurring at the electrode of the last example is

$$Ag(s) + Cl^- \rightleftharpoons AgCl(s) + e$$

Type III – inert metal in contact with solution containing both oxidised and reduced states of an oxidation–reduction system, e.g.

$Pt | Fe^{++}, Fe^{+++}$ reaction at electrode: $\quad Fe^{++} \rightleftharpoons Fe^{+++} + e$

$Au | Sn^{++}, Sn^{++++}$ reaction at electrode: $Sn^{++} \rightleftharpoons Sn^{++++} + 2e$

$Pt | Fe(CN)_6^{----}, \quad$ reaction at electrode: $Fe(CN)_6^{-4} \rightleftharpoons$
$\quad Fe(CN)_6^{---} \qquad\qquad\qquad\qquad\qquad\qquad\qquad Fe(CN)_6^{-3} + e$

According to the convention recommended by the International Union of Pure and Applied Chemistry, Stockholm, 1953, the positive electrode of a cell as written is on the right-hand side. The Daniell cell, written as

$$Zn\,|\,ZnSO_4\,|\,CuSO_4\,|\,Cu$$

has a positive e.m.f. (of 1·0 V) according to this convention. Therefore oxidation takes place at the electrode written on the left-hand side:

$$Zn = Zn^{++} + 2e \tag{15.29}$$

and reduction at the right-hand electrode,

$$Cu^{++} + 2e = Cu \tag{15.30}$$

The net reaction occurring in the cell is

$$Zn + Cu^{++} = Zn^{++} + Cu \tag{15.31}$$

If the sign of the e.m.f. for a cell is not known, then, by the convention, the right-hand electrode as written is considered positive, e.g.

$$Pt,\ H_2\,|\,HCl(m)\,|\,AgCl(s)\,|\,Ag$$

The reaction occurring in this cell is found by adding the reactions at the electrodes:

$$\tfrac{1}{2}H_2 = H^+ + e$$

$$e + AgCl(s) = Cl^- + Ag$$

$$\tfrac{1}{2}H_2 + AgCl(s) = H^+ + Cl^- + Ag \tag{15.32}$$

Had the cell been written as

$$Ag\,|\,AgCl(s)\,|\,HCl(m)\,|\,H_2,\ Pt$$

the reaction occurring in the cell would be written

$$Ag + H^+ + Cl^- = \tfrac{1}{2}H_2 + AgCl(s) \tag{15.33}$$

Which is the correct *spontaneous* reaction can only be determined by measuring the e.m.f. and thus finding the positive electrode. However, it is well known that silver will not dissolve spontaneously in hydrochloric acid to give hydrogen, so that Eq. (15.32) must represent the spontaneous reaction. We can, of course, force the reaction given by Eq. (15.33) to take place by passing a current in a given direction through the cell, i.e. giving the silver electrode a negative potential.

15.12   The standard e.m.f. of reversible cells

Consider a reversible cell in which the reaction

$$aA + bB \rightleftharpoons cC + dD$$

occurs. The change in the Gibbs function for the reaction is given by Eq. (14.6),

$$\Delta G = \Delta G^{\ominus} + RT \ln \frac{a_C^c a_D^d}{a_A^a a_B^b}$$

where the activities are not those when the reaction reaches equilibrium but the activities of the substances as they occur in the cell.

Dividing both sides of this equation by $-nF$, where n is the number of electrons transferred in the electrode reaction, e.g. for reaction (15.31), as seen from (15.29) and (15.30), $n = 2$, or for (15.32) it can be seen that $n = 1$, we have

$$-\frac{\Delta G}{nF} = -\frac{\Delta G^{\ominus}}{nF} - \frac{RT}{nF} \ln \frac{a_C^c a_D^d}{a_A^a a_B^b}$$

Employing Eq. (15.26),

$$E = E^{\ominus} - \frac{RT}{nF} \ln \frac{a_C^c a_D^d}{a_A^a a_B^b} \tag{15.34}$$

where

$$E^{\ominus} = -\frac{\Delta G^{\ominus}}{nF} \tag{15.35}$$

$E^{\ominus}$ is known as the *standard e.m.f. of the cell*. Using the van't Hoff isotherm

$$E^{\ominus} = \frac{RT}{nF} \ln K \tag{15.36}$$

Equation (15.34) is sometimes known as the Nernst equation.

It is seen that if the standard e.m.f. of a cell is known, then the equilibrium constant and the standard Gibbs function change for the cell reaction can be calculated. When the activities of all the reagents are unity, then $E = E^{\ominus}$, i.e. the e.m.f. of the cell when all the substances are in their standard states is the standard e.m.f.

15.13 Determination of standard e.m.f. and mean ionic activity coefficient

As an example, consider a hydrogen electrode and a silver–silver chloride electrode immersed in a solution of hydrochloric acid concentration (m),

$$\text{Pt, } H_2 \, | \, HCl(m) \, | \, AgCl(s) \, | \, Ag$$

The overall reaction for this cell is

$$\tfrac{1}{2}H_2 + AgCl(s) = H^+ + Cl^- + Ag \tag{15.37}$$

From Eq. (15.34), the e.m.f. of the cell is

$$E = E^{\ominus} - \frac{RT}{F} \ln \frac{a_{H^+} \cdot a_{Cl^-} \cdot a_{Ag}}{a_{H_2}^{\frac{1}{2}} \cdot a_{AgCl}} \tag{15.38}$$

We have seen previously, in Section 11.8, that the activities of solids may be put equal to unity. Assuming hydrogen to be an ideal gas, its activity is unity since $P = 101\,325 \text{ N m}^{-2}$. Therefore

$$E = E^{\ominus} - \frac{RT}{F} \ln a_{H^+} \cdot a_{Cl^-} \tag{15.39}$$

Using the mean ionic activity defined by Eq. (15.18), we have

$$E = E^{\ominus} - \frac{2RT}{F} \ln a_{\pm}$$

or

$$E = E^{\ominus} - \frac{2RT}{F} \ln \gamma_{\pm} m$$

Rearranging,

$$E + \frac{2RT}{F} \ln m = E^{\ominus} - \frac{2RT}{F} \ln \gamma_{\pm} \tag{15.40}$$

Real solutions tend to ideal behaviour with dilution, i.e. $\gamma_{\pm} \to 1$ as $m \to 0$. Hence, it is seen from the right-hand side of Eq. (15.40) that the limiting value of $E + [(2RT/F) \ln m]$ is $E^{\ominus}$. That is, if $E + [(2RT/F) \ln m]$ is plotted against m and extrapolated to $m = 0$, then the intercept is $E^{\ominus}$, the e.m.f. of the cell being measured for different acid concentrations.

A better approach is to make use of the Debye–Hückel limiting law,

$$\ln \gamma_{\pm} = -Am^{\frac{1}{2}}$$

Eq. (15.40) becoming

$$E + \frac{2RT}{F} \ln m = E^{\ominus} + \frac{2RTAm^{\frac{1}{2}}}{F} \tag{15.41}$$

If now $E + [(2RT/F) \ln m]$ is plotted against $m^{\frac{1}{2}}$, a straight line graph is obtained for dilute acid solutions, facilitating the extrapolation to determine $E^{\ominus}$. Once this value is known, it may be employed in Eq. (15.40) for the calculation of $\gamma_{\pm}$ for any solution for which the e.m.f. is known. The standard e.m.f. $E^{\ominus}$ of any cell may be calculated from tables of standard electrode potentials (see Section 15.14).

15.14   Standard electrode potential

It would be very convenient if the electrode potentials of half cells were known. From these values the e.m.f.s of all possible combinations of two half-cells to form a reversible cell could be found, which would remove the need to measure the e.m.f. of each of the very large number of possible cells. The e.m.f. of a cell

left-hand half-cell│right-hand half-cell

is
$$E = E_{\text{right}} - E_{\text{left}}. \tag{15.42}$$

It is impossible to measure an absolute electrode potential but we may express all the electrode potentials relative to a reference electrode. The reference electrode, which is assigned the value $E^{\ominus} = 0$ at all temperatures, is taken as the *standard hydrogen electrode*. This is the hydrogen electrode in which hydrogen gas at a pressure of $101\,325$ N m^{-2} (strictly speaking, at a fugacity of $101\,325$ N m^{-2}, but hydrogen is considered an ideal gas) is in contact with platinum immersed in aqueous hydrochloric acid of mean ionic activity $a_{\pm} = 1$. That is,

$$\text{Pt, H}_2(101\,325 \text{ N m}^{-2})│\text{H}^+(a_{\pm} = 1)$$

Consider the cell

$$\text{Pt, H}_2(101\,325 \text{ N m}^{-2})│\text{H}^+(a_{\pm} = 1)│\text{X}^+│\text{X}$$

The e.m.f. of the cell is

$$E = E_{\text{X}} - E_{\text{H}_2}^{\ominus} = E_{\text{X}} \tag{15.43}$$

If the activities of all the substances involved in the cell reaction are unity, then E_X is the standard electrode potential, $E_X^{\ominus}$. For example, consider the cell

$$\text{Pt, H}_2 \,|\, \text{HCl} \,|\, \text{AgCl(s)} \,|\, \text{Ag}$$

The standard e.m.f. of this cell, as determined by the method of Section 15.13, is $E^{\ominus} = 0.2225$ V. Hence, $E_{Ag}^{\ominus} = 0.2225$ V.

Table 15.1 gives a list of some standard electrode potentials.

Table 15.1

Electrode	Reaction	E°V, 298 K
$\text{Li}^+ \,|\, \text{Li}$	$\text{Li}^+ + e \rightarrow \text{Li}$	-3.045
$\text{K}^+ \,|\, \text{K}$	$\text{K}^+ + e \rightarrow \text{K}$	-2.925
$\text{Cs}^+ \,|\, \text{Cs}$	$\text{Cs}^+ + e \rightarrow \text{Cs}$	-2.923
$\text{Na}^+ \,|\, \text{Na}$	$\text{Na}^+ + e \rightarrow \text{Na}$	-2.714
$\text{Al}^{+++} \,|\, \text{Al}$	$\text{Al}^{+++} + 3e \rightarrow \text{Al}$	-1.66
$\text{Zn}^{++} \,|\, \text{Zn}$	$\text{Zn}^{++} + 2e \rightarrow \text{Zn}$	-0.763
$\text{Fe}^{++} \,|\, \text{Fe}$	$\text{Fe}^{++} + 2e \rightarrow \text{Fe}$	-0.440
$\text{Fe}^{+++} \,|\, \text{Fe}$	$\text{Fe}^{+++} + 3e \rightarrow \text{Fe}$	-0.036
$\text{H}^+ \,|\, \text{H}_2, \text{Pt}$	$\text{H}^+ + e \rightarrow \frac{1}{2}\text{H}_2$	0
$\text{Cl}^- \,|\, \text{AgCl(s)} \,|\, \text{Ag}$	$\text{AgCl} + e \rightarrow \text{Cl}^- + \text{Ag}$	$+0.2225$
$\text{Cu}^{++} \,|\, \text{Cu}$	$\text{Cu}^{++} + 2e \rightarrow \text{Cu}$	$+0.337$
$\text{Fe}^{++}, \text{Fe}^{+++} \,|\, \text{Pt}$	$\text{Fe}^{+++} + e \rightarrow \text{Fe}^{++}$	$+0.771$
$\text{Ag}^+ \,|\, \text{Ag}$	$\text{Ag}^+ + e \rightarrow \text{Ag}$	$+0.799$
$\text{Hg}^{++} \,|\, \text{Hg}$	$\text{Hg}^{++} + 2e \rightarrow \text{Hg}$	$+0.854$
$\text{Br}^- \,|\, \text{Br}_2, \text{Pt}$	$\text{Br}_2 + 2e \rightarrow 2\text{Br}^-$	$+1.065$
$\text{Cl}^- \,|\, \text{Cl}_2, \text{Pt}$	$\text{Cl}_2 + 2e \rightarrow 2\text{Cl}^-$	$+1.359$

Due to the fact that the standard hydrogen electrode is so cumbersome and that it may be easily poisoned by gas impurities, it is usual to use other secondary reference electrodes, the standard electrode potentials of which are accurately known. The most common is the calomel electrode $\text{Hg} \,|\, \text{Hg}_2\text{Cl}_2\text{(s)} \,|\, \text{KCl}$.

Consider the cell

$$\text{Zn} \,|\, \text{ZnCl(m)} \,|\, \text{AgCl(s)} \,|\, \text{Ag}$$

The reaction at left electrode is $\qquad \text{Zn} \rightarrow \text{Zn}^{++} + 2e$

The reaction at right electrode is $\quad 2\text{AgCl} + 2e \rightarrow 2\text{Cl}^- + 2\text{Ag}$

Total reaction is $\qquad\qquad\qquad \text{Zn} + 2\text{AgCl} \rightarrow \text{Zn}^{++} + 2\text{Cl}^- + 2\text{Ag}$

From the Nernst equation,

$$E = E^{\ominus} - \frac{RT}{2F} \ln \frac{a_{\text{Zn}^{++}} \cdot a_{\text{Cl}^-}^2 \cdot a_{\text{Ag}}^2}{a_{\text{Zn}} \cdot a_{\text{AgCl}}^2}$$

Hence,

$$E = E_{Ag}^{\ominus} - E_{Zn}^{\ominus} - \frac{RT}{2F} \ln \frac{a_{Zn^{++}}}{a_{Zn}} - \frac{RT}{F} \ln \frac{a_{Cl^-} \cdot a_{Ag}}{a_{AgCl}} \quad (15.44)$$

i.e.

$$E = E_{Ag} - E_{Zn}$$

where

$$E_{Ag} = E_{Ag}^{\ominus} - \frac{RT}{F} \ln \frac{a_{Cl^-} \cdot a_{Ag}}{a_{AgCl}} \quad (15.45)$$

and

$$E_{Zn} = E_{Zn}^{\ominus} - \frac{RT}{2F} \ln \frac{a_{Zn}}{a_{Zn^{++}}} \quad (15.46)$$

E_{Ag} and E_{Zn} being the electrode potentials of the silver and zinc electrodes, respectively.

15.15   The standard Gibbs functions of formation for ions

According to our convention, Pt, $H_2 | H^+(a_{\pm} = 1)$ has $E^{\ominus} = 0$ at all temperatures. Hence, from the equation

$$-\Delta G = nFE$$

the Gibbs function of formation for the hydrogen ion is taken as zero:

$$\tfrac{1}{2}H_2 \longrightarrow H^+ + e \qquad \Delta G^{\ominus} = 0$$

The Gibbs function of formation of an ion relative to this standard may be easily found. For example, for the reversible electrode $Zn^{++} | Zn$, the electrode reaction is

$$Zn^{++} + 2e \longrightarrow Zn \qquad \text{at 298 K, } E^{\ominus} = -0.763 \text{ V}$$

Therefore $\Delta G^{\ominus} = -nFE^{\ominus} = 2F \times 0.763 = 146.8 \text{ kJ mol}^{-1}$

The formation of Zn^{++} is the reverse of this reaction. Hence, the Gibbs function of formation for a zinc ion is $-146.8 \text{ kJ mol}^{-1}$.

15.16   Concentration cells

An example of an electrode concentration cell is a cell comprising two hydrogen electrodes operating at different pressures,

$$\text{Pt, H}_2(\text{fugacity } f_1)\,|\,\text{HCl(a)}\,|\,\text{H}_2(\text{fugacity } f_2),\ \text{Pt}$$

Left-hand reaction $\quad \frac{1}{2}\text{H}_2(f_1) \longrightarrow \text{H}^+(a) + e$

Right-hand reaction $\quad \text{H}^+(a) + e \longrightarrow \frac{1}{2}\text{H}_2(f_2)$

Net reaction $\quad \frac{1}{2}\text{H}_2(f_1) \longrightarrow \frac{1}{2}\text{H}_2(f_2)$

From the Nernst equation,

$$E = -\frac{RT}{2F}\ln\frac{f_2}{f_1} \tag{15.47}$$

The standard e.m.f. $E^{\ominus}$ of a concentration cell is obviously zero (see Eq. 15.42).

Consider the electrolyte concentration cell

$$\text{M}\,|\,\text{M}^+(a_1)\,|\,\text{M}^+(a_2)\,|\,\text{M}$$

If the liquid junction potential is negligible, which is the case for dilute solutions or when a salt bridge is used, the e.m.f. of the cell is

$$E = -\frac{RT}{F}\ln\frac{a_2}{a_1} \tag{15.48}$$

This equation does not apply to cells with transport, when the liquid junction potential must be taken into account.

Consider the cell

$$\text{Pt, H}_2(101\ 325\ \text{N m}^{-2})\,|\,\text{HCl}(a_1)\,|\,\text{HCl}(a_2)\,|\,\text{H}_2(101\ 325\ \text{N m}^{-2}),\ \text{Pt}$$

Left-hand reaction $\quad \frac{1}{2}\text{H}_2 \longrightarrow \text{H}^+(a_{\text{H}^+})_1 + e$

Right-hand reaction $\quad \text{H}^+(a_{\text{H}^+})_2 + e \longrightarrow \frac{1}{2}\text{H}_2$

Net reaction $\quad \text{H}^+(a_{\text{H}^+})_2 \longrightarrow \text{H}^+(a_{\text{H}^+})_1$

The more concentrated electrolyte solution tends to decrease in concentration and the less concentrated to increase. The electricity is carried across the liquid junction by the H^+ and Cl^- ions. Owing to their mobilities being different, they do not transfer the same amount of electricity. *The fraction of the electricity carried by a given ion is known as its transport number.*

Consider the passage of 1 coulomb of electricity through the cell. Then t_+ coulombs of electricity are carried by the H^+ ion and t_- coulombs by the Cl^- ion ($t_+ + t_- = 1$ coulomb).

Suppose that on the left-hand side 1 mole of H^+ ion is formed, then at the liquid junction t_+ mole of H_+ ion passes from left to right.

Net increase (in moles) of H^+ ion concentration $= 1 - t_+ = t_-$.

Also, at the liquid junction t_- mole of Cl^- ion is transferred from right to left.

Now, on the right-hand side, 1 mole of H^+ is removed from solution and at the liquid junction t_+ mole of H^+ ion passes from left to right.

Net decrease (in moles) of H^+ ion concentration $= 1 - t_+ = t_-$.

Also, at the liquid junction t_- mole of Cl^- ion are transferred from right to left.

The net resulting changes in the electrolyte compartments when 1 mole of hydrogen ion is discharged may be written as

$$t_- HCl(a_2) \longrightarrow t_- HCl(a_1)$$

The e.m.f. of the cell is

$$E = -\frac{RT}{F} \ln \frac{a_1^{t_-}}{a_2^{t_-}} = -\frac{t_- RT}{F} \ln \frac{a_1}{a_2} \qquad (15.49)$$

15.17   Determination of solubility product

Consider the cell

$$Ag \,|\, AgCl(s) \,|\, KCl \,\|\, AgNO_3 \,|\, Ag$$

The cell reaction is

$$Ag + Cl^- = AgCl + e$$
$$\underline{Ag^+ + e = Ag}$$
$$Ag^+ + Cl^- = AgCl$$

The standard e.m.f. of the cell, $E^\ominus$, is related to the standard Gibbs function change for the reaction:

$$-\Delta G^\ominus = nFE^\ominus$$

Using the van't Hoff isotherm, we have

$$E^\ominus = \frac{RT}{nF} \ln K$$

we have, for the cell above,

$$E^\ominus = \frac{RT}{F} \ln \frac{a_{Ag^+} \cdot a_{Cl^-}}{a_{AgCl}} = \frac{RT}{F} \ln a_{Ag^+} \cdot a_{Cl^-}$$

i.e.
$$E^\ominus = \frac{RT}{F} \ln (S.P.)$$

Hence, it is seen that a particular solubility product may be determined from a knowledge of the standard electrode potential at a given temperature.

Example   For the cell

$$Pb(s)\,|\,PbCl_2(s)\,|\,HCl(1)\,|\,AgCl(s)\,|\,Ag(s)$$

the standard e.m.f. is 0.490 V at 298 K and the temperature coefficient of the e.m.f. is -1.85×10^{-4} V K^{-1}. Calculate the standard Gibbs function and standard entropy changes for the reaction involved, and also the standard entropy of $PbCl_2(s)$ at 298 K. The standard entropies of Pb, Ag and AgCl are 64.8, 42.5 and 96.0 J K^{-1} mol^{-1}, respectively.

The electrode reactions are

$$Pb + 2Cl^- = PbCl_2 + 2e$$

and

$$2AgCl + 2e = 2Cl^- + 2Ag$$

Net reaction is

$$Pb + 2AgCl(s) = PbCl_2(s) + Ag$$

$$\Delta G^\ominus = -nFE^\ominus$$
$$= -2 \times 96\,487 \times 0.490$$
$$= -94\,350 \text{ J}$$

$$\Delta S^\ominus = nF\left(\frac{\partial E^\ominus}{\partial T}\right)_P$$
$$= 2 \times 96\,487 \times (-1.85 \times 10^{-4})$$
$$= -35.4 \text{ J K}^{-1}$$

Since

$$\Delta S^\ominus = S^\ominus(PbCl_2) + 2S^\ominus(Ag) - \{S^\ominus(Pb) + 2S^\ominus(AgCl)\}$$

we have

$$-35.4 = S^\ominus(PbCl_2) + 2(42.5) - 64.8 - 2(96.0)$$

Hence,

$$S^\ominus(PbCl_2) = 136.4 \text{ J K}^{-1} \text{ mol}^{-1}$$

Problems

$$F = 96\,500 \text{ C}$$

1.   The standard electrode potentials of $Zn|Zn^{++}$ and $Cu|Cu^{++}$ are $+0.763$ V and -0.337 V at 298 K. Calculate the standard Gibbs function change for the reaction $Zn + Cu^{++} = Zn^{++} + Cu$.
Answer -212 kJ

2. The e.m.f. and temperature coefficient of the cell

$$Ag|AgCl(s)|KCl|Hg_2Cl_2(s)|Hg$$

are 0·0456 V and 0·000 339 V K^{-1} at 298 K. Calculate the changes in enthalpy, entropy and the Gibbs function at 298 K for the cell reaction

$$Ag(s)+\tfrac{1}{2}Hg_2Cl_2(s) = AgCl(s)+Hg$$

Answer   5·31 kJ, 32·6 J K^{-1}, $-4·39$ kJ

3. Calculate the e.m.f. and temperature coefficient of the cell

$$Pb|PbCl_2(s)|KCl|AgCl(s)|Ag$$

The changes in the enthalpy and the Gibbs function for the reaction at 298 K occurring in the cell, i.e.

$$Pb+2AgCl(s) = PbCl_2(s)+Ag$$

are $-105·2$ kJ and $-94·5$ kJ.

Answer   0·491 V and $-1·86$ V K^{-1}

4. The electrode potentials of $Ag|Ag^+$ and $Ag|AgBr(s)|Br^-$ are $-0·7994$ V and $-0·0712$ V at 298 K. Calculate the solubility product of AgBr.

Answer $4·78 \times 10^{-13}$

16

Statistical Thermodynamics

16.1 Introduction

The atomic theory of matter and the wave theory of electromagnetic radiation (light) were well accepted and in common use at the beginning of the twentieth century. In 1901 Planck came to the conclusion that light energy was absorbed and emitted in discrete quantities, termed quanta, by means of photons. These energies are integral multiples of hv, where h is Planck's constant and v is the frequency of the radiation, the energy of a photon being

$$E = hv$$

In 1905 Einstein, using the concept of quanta, explained the photoelectric effect. In 1920 de Broglie suggested that matter, and especially electrons, may exhibit wave-like properties. In 1927 Davisson and Gremer's experiments showed that electrons were diffracted by a crystal lattice. At about this time, Schrödinger derived an equation expressing wave-like properties of matter, the solution of which showed that only discrete energy levels were allowed. The lowest energy level of this series is known as the ground level. Schrödinger's work, and also that of Heisenberg on matrix mechanics, led the way to quantum mechanics.

Long before the atomic nature of matter and the quantisation of energy were understood, the foundations of classical thermodynamics were developed. These were based on experience of the behaviour of macroscopic properties.

The microscopic approach advances our understanding of thermodynamics and is therefore extremely valuable. It has to be, of course, a statistical analysis that averages the random, ever-changing characteristics of the particles in a system. The physical data necessary for evaluation purposes include atomic weights, moments of inertia, vibrational frequencies, etc.

Statistical methods may only be applied with ease to simple systems (e.g. monatomic and diatomic gases). In liquids the particles do not act independently, and thus details of the intermolecular forces which must be included complicate the study.

The reader should be familiar with Section 5.13, in which it was shown that

$$S = k \ln W \tag{16.1}$$

and also with such terms as 'microstate' (which requires the description of a system to the ultimate limit of detail), 'macrostate', 'thermodynamic probability' (the number of microstates corresponding to a macrostate of a system) and 'degeneracy'.

Symbolism   The energy of the ground level will be denoted by ε_0, the value of the energy of the first level above the ground level by ε_1.

Consider an isolated system which has energy U and contains N particles. *If* the particles are independent of each other, then

$$U = \sum_{i=0}^{j} n_i \varepsilon_i \tag{16.2}$$

where n_i is the number of particles in the ith energy level and the energy of this level is ε_i. The summation is over all energy levels which contain one or more particles. If the particles are not independent of each other, forces of interaction exist, and thus a term for the energy of interaction must be included.

The total number of particles must be given by

$$N = \sum_{i=0}^{j} n_i \tag{16.3}$$

16.2   Maxwell–Boltzmann, Bose–Einstein and Fermi–Dirac statistics

There are three types of statistics required for the different physical situations found in nature. The main difference between the older Maxwell–Boltzmann statistics and the newer quantum statistics is that the particles are considered indistinguishable in the latter. Also, the Maxwell–Boltzmann statistics, which were developed before the advent of quantum mechanics, are not restricted to the concept of quantisation of energy. This is not important, since the method may easily be adapted to take account of quantised energy states.

In Maxwell–Boltzmann statistics any number of particles may be assigned to an energy level. However, when we deal with electrons and protons, this is not so, since the Pauli exclusion principle requires that only one particle occupy a given energy level.

Fermi–Dirac statistics apply to particles for which the sum of the electrons, protons and neutrons is an odd number (i.e. particles of half integral spin), e.g. electrons(e), protons(H^+), helium3 (He^3), ammonia(NH_4^+), nitric oxide(NO).

Particles such as photons and deuterons, any number of which may occupy the same energy level follow Bose–Einstein statistics, other examples being hydrogen(H_2), deuterium(D_2), nitrogen(N_2), helium4 (He^4). The sum of the number of electrons, protons and neutrons in these particles is even (i.e. they are of integral spin).

16.3   Maxwell–Boltzmann statistics. Any number of particles may be assigned to the same energy level and the particles are distinguishable

For solids, where the particles have fixed positions in space, the particles may be considered as distinguishable. However, for gases, the particles are free to move throughout the entire volume and thus it is impossible to identify individuals. These statistics therefore lead to incorrect results for gases if allowance is not made for the indistinguishability of the particles. All three statistics may be used to determine the properties of ideal gases and lead to the same results.

Consider that there are N distinguishable particles in energy levels ε_0, ε_1, ... The total number of arrangements for placing n_0 particles in the ground level, n_1 in the first level, n_2 in the second level, and so on, is the thermodynamic probability of that macrostate and has been derived previously (page 113):

$$W = N! \prod_i \frac{g_i^{n_i}}{n_i!} \tag{16.4}$$

where g_i is the degeneracy of the ith energy level, i.e. the number of ways the energy ε_i may be achieved. In other words, there are g_i quantum states of energy ε_i.

Energy level $\qquad \varepsilon_0\, \varepsilon_1\, \varepsilon_2\, \cdots\, \varepsilon_i\, \cdots$

Degeneracy $\qquad g_0\, g_1\, g_2\, \cdots\, g_i\, \cdots$

Number of particles $\quad n_0\, n_1\, n_2\, \ldots\, n_i\, \ldots$

16.4   Bose–Einstein statistics. Any number of particles, which are considered to be indistinguishable, may occupy a given energy level

Consider that there are n_i particles in the ith energy level, ε_i, which has a degeneracy g_i. The n_i particles will be distributed among the g_i states. To illustrate and simplify the computation, consider that the ith energy level has $g_i - 1$ partitions and that each of the n_i particles is represented by X. The $g_i - 1$ partitions are just sufficient to separate the energy level into g_i intervals, so that one possible distribution in the energy level may be shown as XX | XXX | | X | ... | X, which shows that there are two particles in the first state, three in the second, none in the third, etc., with one in the last.

The possible number of distributions of n_i particles among the g_i states may be found by permuting the array of partitions and particles. The number of permutations is $(n_i + g_i - 1)!$ However, the partitions and particles are indistinguishable. That is, interchanging two partitions does not alter an arrangement and also interchanging two particles does not alter an arrangement. Therefore, $(n_i + g_i - 1)!$ must be divided by the number of permutations of the n_i particles, $n_i!$, and the number of permutations of the $g_i - 1$ partitions, $(g_i - 1)!$, to find the number of possible arrangements of the n_i particles in level ε_i:

$$\frac{(n_i + g_i - 1)!}{n_i!(g_i - 1)!}$$

The number of ways of distributing N particles among the various energy levels is therefore

$$W = \prod_i \frac{(n_i + g_i - 1)!}{n_i!(g_i - 1)!} \tag{16.5}$$

16.5   Fermi–Dirac statistics. Only one particle may occupy a given energy level, all particles being indistinguishable

Consider that the ith energy level has degeneracy g_i and that n_i particles are distributed among these g_i states ($n_i < g_i$). To assist us, let us consider that the particles are distinguishable. Then the first particle may be placed in any one of the g_i states and for each one of these choices the second particle may be placed in any one of the remaining $g_i - 1$ states, and so on. Therefore the number of arrangements is

$$g_i(g_i - 1) \dots (g_i - n_i + 1) = \frac{g_i!}{(g_i - n_i)!}$$

But since the particles are in fact indistinguishable, we must divide by the possible number of permutations of n_i particles, $n_i!$. Therefore the number of arrangements of n_i particles in the ith level is

$$\frac{g_i!}{n_i!(g_i - n_i)!}$$

The number of ways of distributing N particles among the various levels is

$$W = \prod_i \frac{g_i!}{n_i!(g_i - n_i)!} \tag{16.6}$$

16.6 The Maxwell–Boltzmann distribution

It has been seen previously that the entropy and thermodynamic probability of any state of a system are related:

$$S = k \ln W$$

However, for an equilibrium state, the probability must be a maximum. Hence, for an equilibrium state,

$$S = k \ln W_{max} \tag{16.7}$$

We shall now find the distribution which will make W a maximum. In this section, we shall consider the distribution of distinguishable particles, any number of which may be assigned to the same level, i.e. Maxwell–Boltzmann statistics.

We wish to maximise the function W (Eq. 16.4). In fact, it is more convenient to maximise the logarithm of the probability. (In addition, this may then be employed directly to calculate entropy.) From calculus, it is known that a maximum may be found by equating the derivative to zero. That is, at equilibrium,

$$d \ln W = \frac{\partial \ln W}{\partial n_1} \cdot dn_1 + \frac{\partial \ln W}{\partial n_2} dn_2 + \dots + \frac{\partial \ln W}{\partial n_i} dn_i + \dots$$

$$= \sum_i \frac{\partial \ln W}{\partial n_i} dn_i = 0 \qquad (16.8)$$

For simplicitly, we shall deal with a *closed* system of *independent* particles. That is, (1) the total number of particles, N, is constant, $N = \Sigma_i n_i$ (16.9); (2) the total energy, U, is constant, $U = \Sigma_i n_i \varepsilon_i$ (16.10). Since, at equilibrium, the values of the energy levels will not alter, (i.e. $d\varepsilon_i = 0$) we have, from Eq. (16.9), that

$$\Sigma_i \, dn_i = 0 \qquad (16.11)$$

and, from Eq. (16.10), that

$$\Sigma_i \varepsilon_i \, dn_i = 0 \qquad (16.12)$$

From Eq. (16.4), we have that

$$\ln W = \ln N! + \Sigma_i n_i \ln g_i - \Sigma_i \ln n_i!$$

Now Stirling's formula gives a good approximation to $\ln x!$ where x is a large·number:

$$\ln x! \doteqdot x \ln x - x \qquad (16.13)$$

Employing this to substitute for $\ln n_i!$,

$$\ln W = \ln N! + \Sigma_i n_i \ln g_i - \Sigma_i n_i \ln n_i + \Sigma_i n_i$$

Differentiating, remembering that N and g_i are constants, and using Eq. (16.11):

$$d \ln W = \Sigma_i \ln g_i \cdot dn_i - \Sigma_i \ln n_i \cdot dn_i - \Sigma_i n_i \, d \ln n_i$$

But $$\Sigma_i n_i \, d \ln n_i = \Sigma_i n_i \frac{dn_i}{n_i} = \Sigma_i \, dn_i = 0$$

Therefore, at equilibrium,

$$d \ln W = \Sigma_i \ln g_i \cdot dn_i - \Sigma_i \ln n_i \cdot dn_i = 0 \qquad (16.14)$$

This equation gives the change in $\ln W$ as the number of particles in each level is varied.

If the variations in each of the n_i were *independent* of one another — in other words, if the system were open so that each n_i could vary without restriction — then the solution of Eq. (16.8) would be found by *placing each of the coefficients of the* dn_i *terms equal to zero*. However, since the total number of particles is constant, the values of the dn_i are not independent of each other (Eq. 16.11). Also, the energy of the system is constant. Our problem is to maximise $\ln W$, i.e. find the solution of d ln $W = 0$, under two constraints (Eqs 16.11 and 16.12).

One method for the solution of such a problem is that of *Lagrangian undetermined multipliers.*

From Eq. (16.14), we have

$$\Sigma_i \ln \frac{g_i}{n_i} \cdot dn_i = 0 \qquad (16.15)$$

Multiplying Eqs (16.11) and (16.12) by arbitrary constants α and β (known as Lagrangian multipliers), and subtracting the results from Eq. (16.15), we have

$$\Sigma_i \left[\ln \frac{g_i}{n_i} - \alpha - \beta\varepsilon_i \right] dn_i = 0 \qquad (16.16)$$

It is possible to select values of α and β such that one of the terms in the sum (e.g. $i = 1$) is zero, the value of dn_1 being immaterial. The remaining dn_i terms are then independent of one another (since dn_1 may be obtained from these dn_i terms (Eq. 16.11). It is now possible to set each of the coefficients of dn_i to zero in Eq. (16.16):

$$\ln \frac{g_i}{n_i} - \alpha - \beta\varepsilon_i = 0 \qquad (16.17)$$

Hence, the value of n_i for energy ε_i which leads to a maximum in W (i.e. which gives the distribution for the most probable macrostate) is

$$\boxed{n_i = g_i e^{-\alpha - \beta\varepsilon_i}} \qquad (16.18)$$

We must now identify the two constants α and β in this distribution equation.

16.7 Identification of α

Now we have

$$\Sigma_i n_i = N$$

Thus, from Eq. (16.18),

$$\Sigma_i g_i e^{-\alpha - \beta\varepsilon_i} = N$$

Therefore

$$e^{-\alpha} = \frac{N}{\Sigma_i g_i e^{-\beta\varepsilon_i}}$$

or
$$e^{-\alpha} = \frac{N}{Z}$$
(16.19)

where
$$Z = \Sigma_i \, g_i \, e^{-\beta \varepsilon_i}$$
(16.20)

Z is known as the molecular partition function and is of the utmost importance in statistical thermodynamics.

The Maxwell–Boltzmann distribution, Eq. (16.18), is therefore

$$n_i = \frac{N g_i \, e^{-\beta \varepsilon_i}}{Z}$$
(16.21)

16.8   Identification of β

It will be shown that the constant β may be associated with the reciprocal of the macroscopic property T.

The Maxwell–Boltzmann thermodynamic probability is

$$W = N! \prod_i \frac{g_i^{n_i}}{n_i!}$$

Taking logarithms,

$$\ln W = \ln N! + \Sigma_i (n_i \ln g_i - \ln n_i!)$$

Applying Stirling's approximation (Eq. 16.13) to $\ln N!$ and $\ln n_i!$, we have

$$\ln W = N \ln N - N + \Sigma_i \, (n_i \ln g_i - n_i \ln n_i + n_i)$$
$$= N \ln N + \Sigma_i \, n_i \ln g_i - \Sigma_i \, n_i \ln n_i$$
(16.22)

On taking logarithms, Eq. (16.21) gives

$$\ln n_i = \ln N - \ln Z + \ln g_i - \beta \varepsilon_i$$

Substituting into Eq. (16.22),

$$\ln W = N \ln N + \Sigma_i \, n_i \ln g_i - \Sigma_i \, n_i (\ln N - \ln Z + \ln g_i - \beta \varepsilon_i)$$
$$= N \ln N + \Sigma_i \, n_i \ln g_i - N \ln N + N \ln Z$$
$$- \Sigma_i \, n_i \ln g_i + \beta \Sigma_i \, n_i \varepsilon_i$$
$$\ln W = N \ln Z + \beta U$$

Using Eq. (16.1),

$$S = kN \ln Z + k\beta U$$
(16.23)

(This equation applies to a system of distinguishable particles only, since it is based on Maxwell–Boltzmann statistics.)

From the classical thermodynamics of a simple system, Eq. (5.16) states that, at constant volume, $T\mathrm{d}S = \mathrm{d}U$. That is,

$$\left(\frac{\partial S}{\partial U}\right)_V = \frac{1}{T} \tag{16.24}$$

Let us now differentiate Eq. (16.23) with respect to U at constant V:

$$\left(\frac{\partial S}{\partial U}\right)_V = \frac{Nk}{Z} \cdot \frac{\mathrm{d}Z}{\mathrm{d}\beta}\left(\frac{\partial \beta}{\partial U}\right)_V + \beta k + kU\left(\frac{\partial \beta}{\partial U}\right)_V \tag{16.25}$$

From Eq. (16.20),

$$\frac{\mathrm{d}Z}{\mathrm{d}\beta} = -\frac{UZ}{N}$$

Hence, the first and last terms in Eq. (16.25) cancel, giving

$$\left(\frac{\partial S}{\partial U}\right)_V = \beta k$$

On comparing this equation with Eq. (16.24), it is seen that

$$\boxed{\beta = \frac{1}{kT}} \tag{16.26}$$

The Maxwell–Boltzmann distribution law is therefore

$$\boxed{n_i = \frac{Ng_i e^{-\varepsilon_i/kT}}{Z}} \tag{16.27}$$

and the molecular partition function is given by

$$\boxed{Z = \Sigma_i g_i e^{-\varepsilon_i/kT}} \tag{16.28}$$

16.9 Comparison of the three types of statistics

In Section 16.6, the Maxwell–Boltzmann distribution law was derived. In an exactly similar manner the distribution laws for Bose–Einstein and Fermi–Dirac statistics may be obtained, and these are given in Table 16.1.

In Figures 16.1, 16.2 and 16.3, the number of particles per energy level is plotted against the energy for various given temperatures for the three types of statistics.

The constant α may be determined for each distribution in a similar manner to that of Section 16.7. Also, note that the constant

Table 16.1

	W	n_i
Maxwell–Boltzmann	$W_{MB} = N! \, \Pi_i \dfrac{g_i^{n_i}}{n_i!}$	$n_i = \dfrac{g_i}{e^{\alpha + \varepsilon_i/kT}}$
Bose–Einstein	$W_{BE} = \Pi_i \dfrac{(n_i + g_i - 1)!}{n_i!(g_i - 1)!}$	$n_i = \dfrac{g_i}{e^{\alpha + \varepsilon_i/kT} - 1}$
Fermi–Dirac	$W_{FD} = \Pi_i \dfrac{g_i!}{n_i!(g_i - n_i)!}$	$n_i = \dfrac{g_i}{e^{\alpha + \varepsilon_i/kT} + 1}$

β is, in all three statistics, equal to $1/kT$ and this may again be shown in a similar manner to that of Section 16.8.

When the exponential term is very large compared with unity (i.e. g_i/n_i is very large compared with unity, meaning that the degeneracy of each level is large compared with the number of particles in the level), then the distributions according to the newer statistics reduce to the classical Maxwell–Boltzmann distribution.

For all real gases, the Maxwell–Boltzmann distribution gives accurate results, exceptions being hydrogen and helium at low temperatures, to which Bose–Einstein statistics apply.

As previously mentioned, Fermi–Dirac and Bose–Einstein statistics apply to indistinguishable particles:

$$W_{FD} = \prod_i \frac{g_i!}{n_i!(g_i - n_i)!}$$

i.e.

$$W_{FD} = \prod_i \frac{g_i(g_i - 1)\ldots(g_i - n_i + 1)}{n!}$$

$$W_{BE} = \prod_i \frac{(g_i + n_i - 1)!}{n_i!(g_i - n_i)!}$$

i.e.

$$W_{BE} = \prod_i \frac{g_i(g_i + 1)\ldots(g_i + n_i - 1)}{n_i!}$$

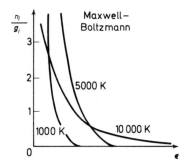

Figure 16.1. Maxwell–Boltzmann distribution

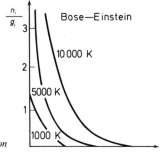

Figure 16.2. Bose–Einstein distribution

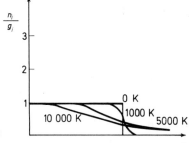

Figure 16.3. Fermi–Dirac distribution

For the special case when $g_i \gg n_i$, we have

$$W_{FD} = W_{BE} = \prod_i \frac{g_i^{n_i}}{n_i!} \qquad (16.29)$$

Note that, of course, the Maxwell–Boltzmann probability W_{MB}, which was computed for distinguishable particles, also reduces to Eq. (16.29) when the particles are considered indistinguishable. This can be seen to be so if we remember that interchanging two indistinguishable particles does not alter a state. Since there are $N!$ ways of permuting N particles, W_{MB} would have to be divided by $N!$.

16.10   Microscopic analogues of heat and work: concept of a reversible process

We shall consider the effect of reversible heat transfer at constant volume on a simple system of independent particles.

Classically, at constant volume,

$$đq = dU$$

Microscopically,

$$U = \Sigma_i n_i \varepsilon_i$$

Hence,

$$dU = \Sigma_i n_i d\varepsilon_i + \Sigma_i \varepsilon_i \, dn_i \qquad (16.30)$$

As might be expected for a system of independent particles, it can be shown by quantum mechanics that the energy levels of the system remain fixed during a heat transfer at constant volume, i.e. the energy of the system is increased by particles being excited to higher energy levels. Hence, the first summation in Eq. (16.30) is zero, so that

$$đq = \Sigma_i \varepsilon_i \, dn_i \qquad (16.31)$$

For a closed system, $\Sigma_i \, dn_i = 0$ — that is, the particles are redistributed among the energy levels such that a maximum probability under the new conditions is attained at equilibrium.

Equation (16.30) is a general equation in which is embodied the effect of reversible heat and work transfer. Hence, with Eq. (2.5) in mind,

$$đw = -\Sigma_i n_i \, d\varepsilon_i \qquad (16.32)$$

Note that in the absence of heat transfer, the n_i values remain constant. That is, during a reversible work transfer the energy levels change in position but the distribution among these new levels is not altered.

Equation (16.30) may be regarded as a microscopic statement of the first law of thermodynamics. It is obviously possible to have a process in which the values of n_i and ε_i alter. However, if the process is reversible, then only heat absorbed across the system boundaries will alter the distribution, i.e. n_i (Eq. 16.31), and only work transfer will alter the position of the energy levels (Eq. 16.32). If a process is irreversible, both n_i and ε_i may alter. The values of n_i can only be determined for a reversible process. This is possible if the distribution function (Table 16.1) is used; which is derived for an equilibrium state. That is, if the process consists of a series of equilibrium states, i.e. it is reversible, then the state of the system at any stage is given by the most probable distribution. There is no way of knowing the state of a system during an irreversible process, since the distributions are only calculated for equilibrium states.

16.11 Relation of the partition function to the properties of an ideal gas system

For ideal gases, the newer statistics apply. However, the Maxwell–Boltzmann statistics may also be applied, provided that the indistinguishability of the particles is taken into account.

Internal energy Since we are dealing with a system of independent molecules, the internal energy is

$$U = \Sigma_i n_i \varepsilon_i$$

Except for extremely low temperatures, the thermodynamic probability is

$$W = \prod_i \frac{g_i^{n_i}}{n_i!} \tag{16.33}$$

since the degeneracy of each energy level is large compared with the number of molecules occupying the level (see Section 16.9). The degeneracy of any level is large if the differences between the accessible translational, rotational, vibrational and electronic energy levels are small compared with the energy to be distributed.

The distribution approximates to that of Maxwell–Boltzmann,

$$U = \frac{N\Sigma_i g_i \varepsilon_i e^{-\varepsilon_i/kT}}{Z} \tag{16.34}$$

From Eq. (16.28), the differential of the logarithm of Z with respect to $1/T$ is

$$\left(\frac{\partial \ln Z}{\partial 1/T}\right)_V = -\frac{\Sigma_i g_i \varepsilon_i e^{-\varepsilon_i/kT}}{kZ} \tag{16.35}$$

Comparing Eq. (16.35) with Eq. (16.34),

$$U = -Nk\left(\frac{\partial \ln Z}{\partial 1/T}\right)_V = NkT^2\left(\frac{\partial \ln Z}{\partial T}\right)_V$$

For a system consisting of one mole, N is equal to Avogadro's constant L, hence

$$\boxed{u = RT^2\left(\frac{\partial \ln Z}{\partial T}\right)_V} \tag{16.36}$$

Enthalpy $H = U + PV$

For 1 mole, $h = u + Pv$

For an ideal gas, $Pv = RT$

Therefore $$h = RT^2\left(\frac{\partial \ln Z}{\partial T}\right)_V + RT \tag{16.37}$$

Molar heat capacity   At constant volume this is given by

$$c_V = \left(\frac{\partial U}{\partial T}\right)_V = \frac{\partial}{\partial T}\left(RT^2\frac{\partial \ln Z}{\partial T}\right)_V$$

Therefore, $$c_V = \frac{R}{T^2}\left[\frac{\partial^2 \ln Z}{\partial\left(\frac{1}{T}\right)^2}\right]_V \tag{16.38}$$

Entropy   Given by

$$S = k \ln W = k\Sigma_i \ln \frac{g_i^{n_i}}{n_i!}$$

$$= k(\Sigma_i n_i \ln g_i - \Sigma_i \ln n_i!)$$

for a system of indistinguishable particles, since it is based on Eq. (16.33). Using Stirling's formula,

$$S = k\Sigma_i n_i \ln \frac{g_i}{n_i} + kN \tag{16.39}$$

From the distribution law (Eq. 16.27),

$$\ln \frac{g_i}{n_i} = \ln \frac{Z}{N} + \frac{\varepsilon_i}{kT}$$

Substituting this into Eq. (16.39),

$$S = k\Sigma_i n_i \ln \frac{Z}{N} + k\Sigma_i \frac{n_i \varepsilon_i}{kT} + kN$$

Therefore

$$S = kN \ln \frac{Z}{N} + \frac{U}{T} + kN$$

For 1 mole of an ideal gas,

$$\boxed{s = R \ln \frac{Z}{L} + \frac{u}{T} + R} \qquad (16.40)$$

The Helmholtz function

$$A = U - TS$$

For 1 mole,

$$a = u - RT \ln \frac{Z}{L} - u - RT$$

i.e.

$$\boxed{a = -RT \ln \frac{Z}{L} - RT} \qquad (16.41)$$

Pressure
From Eq. (8.16),

$$P = -\left(\frac{\partial A}{\partial V}\right)_T$$

Hence,

$$P = RT \left(\frac{\partial \ln Z}{\partial V}\right)_T \qquad (16.42)$$

The Gibbs function

$$G = H - TS = A + PV$$

For 1 mole of an ideal gas,

$$Pv = RT$$

so that

$$g = a + RT$$

Hence,

$$g = -RT \ln \frac{Z}{L} \qquad (16.43)$$

It can be seen that, once the molecular partition function for a given state is known, all the thermodynamic properties of a system for that state may be calculated.

16.12   Evaluation of the molecular partition function for an ideal gas

In order to carry out the evaluation of the molecular partition function, we must first compute the values of the energy levels. These may be obtained from solutions of the Schrödinger wave equation.

The total energy ε of a molecule is composed of translational energy, rotational energy, vibrational energy and electronic energy. A simplifying assumption is that the energy of a molecule is the sum of these energies:

$$\varepsilon = \varepsilon_{trans} + \varepsilon_{rot} + \varepsilon_{vib} + \varepsilon_{elec} \qquad (16.44)$$

In other words, we assume that no interaction terms are necessary, e.g. rotational motion is not affected by vibrational motion.

The concept of separable energy states allows the Schrödinger wave equation to be split into a number of differential equations which are easier to solve.

It follows from Eq. (16.28) that the partition function is given by

$$Z = Z_{trans} \cdot Z_{rot} \cdot Z_{vib} \cdot Z_{elec} \qquad (16.45)$$

16.13   The translational partition function

The wave equation for the translational motion of a particle in a box is

$$\frac{\partial^2 \psi}{\partial x^2} + \frac{\partial^2 \psi}{\partial y^2} + \frac{\partial^2 \psi}{\partial z^2} = -\frac{8\pi^2 m}{h^2} \varepsilon_{trans} \psi$$

where it is assumed that the particle moves in the absence of a potential field, and thus its potential energy is arbitrarily set to zero. Further simplification of this equation is obtained by writing

$$\psi = \psi_x . \psi_y . \psi_z$$

where ψ_x is a function of x only, similarly ψ_y and ψ_z, the total translational energy being

$$\varepsilon_{trans} = \varepsilon_x + \varepsilon_y + \varepsilon_z \tag{16.46}$$

Therefore we have

$$\frac{d^2\psi_x}{dx^2} = -\frac{8\pi^2 m}{h^2}\varepsilon_x\psi_x$$

In solving this equation for ψ_x, in which we are not particularly interested, it is found that

$$\varepsilon_x = \frac{n_x^2 h^2}{8ma^2} \tag{16.47}$$

where a is the length of the box in the x direction, and n_x is a positive integer known as the translational quantum number in the x direction.

Similar equations are obtained for the y and z directions. Quantum mechanics shows that the degeneracy of each level is unity.

The translational energy levels are determined by the quantum numbers n_x, n_y, n_z,

$$\varepsilon_{trans} = \frac{h^2}{8m}\left(\frac{n_x^2}{a^2} + \frac{n_y^2}{b^2} + \frac{n_z^2}{c^2}\right) \tag{16.48}$$

The energy of the ground level is extremely small, of the order of 10^{-18} J mol^{-1}, compared with the average translational energy of a gas at 300 K (as calculated from kinetic theory), which is about 3 500 J mol^{-1}. Also, the differences between neighbouring energy levels are so very small that we may regard translational energies as being continuous and not discrete.

From Eq. (16.28) the translational partition function is given by

$$Z_{trans} = \Sigma_{n_x}\Sigma_{n_y}\Sigma_{n_z} \exp\left[-\frac{h^2}{8mkT}\left(\frac{n_x^2}{a^2} + \frac{n_y^2}{b^2} + \frac{n_z^2}{c^2}\right)\right]$$

This equation may be factorised:

$$Z_{trans} = Z_x . Z_y . Z_z \tag{16.49}$$

where

$$Z_x = \Sigma_{n_x} e^{-An^2}$$

and

$$A = \frac{h^2}{8mkTa^2}$$

Considering helium gas, for which $m = 6\cdot6 \times 10^{-27}$ kg, at 300 K between walls 1 cm apart, and inserting the values of h $(6\cdot626 \times 10^{-34}$ J s$)$ and k $(1\cdot380 \times 10^{-23}$ J K$^{-1})$, the constant A is found to be of the order 10^{-17}.

The first term in the summation, $n_x = 1$, is practically unity and

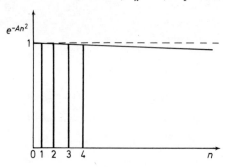

Figure 16.4.

so is the second term and so on. The terms diminish so slowly that the summation may be replaced by an integration,

$$Z_x = \sum_{n_x = 1}^{\infty} e^{-An_x^2} = \int_0^{\infty} e^{-An_x^2} \, dn_x$$

This is a standard integral, giving

$$Z_x = \tfrac{1}{2}\left(\frac{\pi}{A}\right)^{\frac{1}{2}}$$

i.e.

$$Z_x = \frac{a}{h}(2\pi mkT)^{\frac{1}{2}}$$

Similar results are obtained for Z_y and Z_z. Hence,

$$Z_{\text{trans}} = Z_x \cdot Z_y \cdot Z_z = \frac{abc(2\pi mkT)^{\frac{3}{2}}}{h^3}$$

abc is the volume of the system, so that

$$Z_{\text{trans}} = \frac{V}{h^3}(2\pi mkT)^{\frac{3}{2}} \qquad (16.50)$$

If the amount of gas is 1 mole, then V is the molar volume v and m is the molecular weight in kg.

The translational partition function for a given species is seen to be a function of the temperature and volume of the system. All molecules, polyatomic or monatomic, possess translational energy and the contribution to the partition function is given by Eq. (16.50).

16.14 The rotational partition function

In solving the Schrödinger equation for diatomic or linear polyatomic molecules, assuming they behave as rigid rotators (i.e. the interatomic distances are fixed), 'it is found that the permissible energy levels are given by

$$\varepsilon_{\text{rot}} = \frac{J(J+1)h^2}{8\pi^2 I} \tag{16.51}$$

where J is the rotational quantum number, which may have values 0, 1, 2, 3, ..., and I is the moment of inertia of the molecules (about an axis perpendicular to that of the moleculor/axis.

It is also found that each rotational energy level has a degeneracy of $2J+1$. From Eq. (16.28), the rotational partition function is therefore given by

$$Z_{\text{rot}} = \sum_{J=0}^{\infty} (2J+1)\exp-\left(\frac{J(J+1)h^2}{8\pi^2 IkT}\right) = \sum_{J=0}^{\infty} (2J+1)\exp[-J(J+1)B] \tag{16.52}$$

where $B = h^2/8\pi^2 IkT$.

Although the spacing between neighbouring rotational energy levels is much larger than that for translational levels, if $kT \gg h^2/8\pi^2 I$, the rotational levels will be reasonably close together compared with the translational energy levels, and thus it is again permissible to replace the summation by an integration. (*At low temperatures*, the inequality is not valid and the rotational partition function *must* be found *by summation* (Eq. 16.52.)

$$Z_{\text{rot}} = \int_0^{\infty} (2J+1)\,e^{-J(J+1)B}\,dJ$$

Noting that $(2J+1)\,dJ = d(J^2+J)$, we may write

$$Z_{\text{rot}} = \int_0^{\infty} e^{-(J^2+J)B}\,d(J^2+J)$$

$$= -\left[\frac{e^{-(J^2+J)B}}{B}\right]_0^{\infty} = \frac{1}{B}$$

Hence,

$$Z_{\text{rot}} = \frac{8\pi^2 IkT}{h^2} \tag{16.53}$$

This equation is valid for a diatomic molecule which is hetero-nuclear (i.e. which has different atoms), e.g. CO, HD, HCl, $^{35}Cl^{37}Cl$, or for an unsymmetrical linear polyatomic molecule which behaves as a rigid rotator, assuming a continuum of energy.

For diatomic molecules which are homonuclear (i.e. have identical atoms), e.g. H_2, N_2, O_2, and linear polyatomic symmetrical molecules, e.g. $O=C=O$, two rotations which would otherwise be regarded as different become equivalent. For molecules with indistinguishable orientations, a *symmetry number* σ must be introduced to the partition function to avoid too many orientations being taken into account,

$$Z_{rot} = \frac{8\pi^2 IkT}{\sigma h^2} \qquad (16.54)$$

where $\sigma = 2$ for symmetrical molecules and $\sigma = 1$ for non-symmetrical molecules.

In addition to Eq. (16.54) not being valid at low temperatures, owing to the assumption that $kT \gg h^2/8\pi^2 I$, the rigid rotator model is, in practice, not valid at high temperatures. For non-linear molecules, it can be shown that

$$Z_{rot} = \frac{8\pi^2(8\pi^3 I_x I_y I_z)^{\frac{1}{2}}(kT)^{\frac{3}{2}}}{\sigma h^3} \qquad (16.55)$$

where I_x, I_y, I_z are the three principal moments of inertia of the molecule. The symmetry number σ is equal to the number of indistinguishable orientations of the molecule, e.g. for H_2O, $\sigma = 2$; for NH_3, $\sigma = 3$; for CH_4, $\sigma = 12$.

The rotational partition function may be calculated directly from Eq. (16.28) by substituting the values of the rotational energy levels which may be obtained from spectroscopic measurement.

$$Z_{rot} = g_0 e^{-\varepsilon_0/kT} + g_1 e^{-\varepsilon_1/kT} + g_2 e^{-\varepsilon_2/kT} + \ldots \qquad (16.56)$$

where ε_0, ε_1, $\varepsilon_2, \ldots$ are the values of the rotational energy levels (ground, first, second, etc). The ground level is usually taken to have zero energy.

16.15   Internal rotation

For molecules such as CH_3-CH_3, internal rotation of the methyl groups along the carbon–carbon bond is possible, whereas for

$CH_2=CH_2$, for example, a torsional, twisting vibration is possible, but not complete rotation. The internal rotational partition function is usually complicated by the presence of potential energy barriers which must be overcome for free rotation. Hence, no simple expression exists for this particular function.

16.16 The vibrational partition function

A diatomic molecule may be treated to a first approximation as a harmonic oscillator, the Schrödinger equation being

$$\frac{d^2\psi}{dx^2} + \frac{8\pi^2 m}{h^2}(\varepsilon_{vib} - \tfrac{1}{2}kx^2)\psi = 0$$

where k is the force constant. On solving this equation, it is found that

$$\varepsilon_{vib} = (v+\tfrac{1}{2})hv \qquad (16.57)$$

where v is the vibrational quantum number, $v = 0, 1, 2, ...$, and v is the fundamental frequency of vibration. The energy levels are non-degenerate. Even for the ground level $v = 0$, the molecule has vibrational energy $\tfrac{1}{2}hv$.

We have

$$Z_{vib} = \sum_{v=0}^{\infty} e^{\frac{-(v+\frac{1}{2})hv}{kT}} = e^{-hv/2kT} \sum_{v=0}^{\infty} e^{-vhv/kT} \qquad (16.58)$$

The energy difference between vibrational levels (equally spaced by hv) is extremely large compared with the rotational or translational levels, i.e. $hv > kT$. Hence, the summation cannot be replaced by integration. However, the summation may be reduced to a simpler form. Putting $x = hv/kT$, we have

$$\sum_{v=0}^{\infty} e^{-vx} = 1 + e^{-x} + e^{-2x} + e^{-3x} + ... = (1-e^{-x})^{-1}$$

Hence, from Eq. (16.58),

$$\boxed{Z_{vib} = \frac{e^{-hv/2kT}}{1-e^{-hv/kT}}} \qquad (16.59)$$

The fundamental vibrational frequency may be found by spectroscopic measurement and thus used in Eq. (16.59). Alternatively, the experimental values of the vibrational energy levels, which are obtained from spectra, may be employed in Eq. (16.28) to calculate the vibrational partition function.

The partition function relative to the ground level is

$$Z_{vib} = (1 - e^{-x})^{-1} \tag{16.60}$$

Remembering that the levels are non-degenerate, we have

$$Z_{vib} = e^{-\varepsilon_0/kT} + e^{-\varepsilon_1/kT} + e^{-\varepsilon_2/kT} + \ldots \tag{16.61}$$

This is, of course, the most accurate way of calculating the partition function, since it allows for the anharmonicity of the vibration.

As mentioned previously, the spacing of vibrational energy levels is large compared with kT, and thus all but the first term in the summation may be neglected. If, as usual, ε_0 is arbitrarily set to zero, then $Z_{vib} = 1$. (Allowance for ε_0 actually being equal to $\frac{1}{2}h\nu$ can be made.)

16.17 Electronic partition function

The electronic partition function may be calculated from Eq. (16.28), employing spectroscopic data. However, the spacing between electronic levels is usually so large compared with kT that, at normal temperatures, only the ground level is occupied. For example, for atomic hydrogen, the first level, ε_1, is about 160×10^{-20} J above the ground level. The value of kT at $300 K$ is 0.41×10^{-20} J. Since

$$Z_{elec} = g_0 e^0 + g_1 e^{-\varepsilon_1/kT} + g_2 e^{-\varepsilon_2/kT} + \ldots$$

all but the first term will be negligible. Therefore, in general,

$$\boxed{Z_{elec} = g_0} \tag{16.62}$$

For atomic hydrogen, $g_0 = 2$ since the electron may have spin quantum numbers $\pm\frac{1}{2}$, i.e. the ground level degeneracy is 2. Therefore, for hydrogen, $Z_{elec} = 2$. For helium, the first shell is complete; thus there is no degeneracy. For all inert gases, in fact, the electronic energy levels of the ground state are completely filled. Hence, the ground level is non-degenerate, so that $Z_{elec} = 1$. For the alkali metals, the outermost electron may have $\frac{1}{2}$ spin since that level is not filled completely. Thus $Z_{elec} = 2$. In most covalent molecules the ground level is completely filled, so that $Z_{elec} = 1$. A notable exception is O_2, for which $g_0 = 3$; thus $Z_{elec} = 3$.

16.18   Properties of a monatomic ideal gas

A monatomic gas can only have translational and electronic energies. That is, $Z = Z_{trans} \cdot Z_{elec}$. Since $Z_{elec} = g_0$, where g_0 is the degeneracy of the electronic ground level, we have

$$Z = g_0 \frac{v}{h^3} (2\pi mkT)^{\frac{3}{2}}$$ (16.63)

From Eq. (16.36), the molar internal energy is

$$u = RT^2 \left(\frac{\partial \ln Z}{\partial T} \right)_V$$

The partial derivative of $\ln Z$, on the basis of Eq. (16.63), is $\frac{3}{2} T^{-1}$; thus

$$u = \frac{3}{2} RT$$ (16.64)

The molar heat capacity at constant volume is

$$c_V = \left(\frac{\partial U}{\partial T} \right)_V = \frac{3}{2} R$$

The pressure is given by Eq. (16.42) — that is,

$$P = RT \left(\frac{\partial \ln Z}{\partial V} \right)_T = \frac{RT}{v}$$

which is, of course, the equation of state for 1 mole of ideal gas. The molar enthalpy is

$$h = u + Pv = \frac{3}{2} RT + RT = \frac{5}{2} RT$$

The molar entropy, as given by Eq. (16.40), is

$$s = R \ln \frac{Z}{L} + \frac{u}{T} + R$$

Hence,

$$s = R \ln \left[\frac{v g_0 (2\pi mkT)^{\frac{3}{2}}}{L h^3} \right] + \frac{5}{2} R$$ (16.65)

(NOTE: m is the molecular weight in kg.) This is known as the *Sackur–Tetrode* equation. The molar Helmholtz function is given by

$$a = u - Ts = \frac{3}{2}RT - RT \ln\left[\frac{vg_0(2\pi mkT)^{\frac{3}{2}}}{Lh^3}\right] - \frac{5}{2}RT$$

Hence,

$$a = -RT - RT \ln\left[\frac{vg_0(2\pi mkT)^{\frac{3}{2}}}{Lh^3}\right]$$

The molar Gibbs function is given by

$$g = h - Ts = \frac{5}{2}RT - RT \ln\left[\frac{vg_0(2\pi mkT)^{\frac{3}{2}}}{Lh^3}\right] - \frac{5}{2}RT$$

Hence,

$$g = -RT \ln\left[\frac{vg_0(2\pi mkT)^{\frac{3}{2}}}{Lh^3}\right]$$

16.19   Properties of a diatomic ideal gas

A diatomic ideal gas may have translational, rotational, vibrational and electronic energies, the partition function being given by

$$Z = Z_{\text{trans}} \cdot Z_{\text{rot}} \cdot Z_{\text{vib}} \cdot Z_{\text{elec}}$$

The translational and electronic contributions have already been calculated in considering the properties of a monatomic ideal gas. We therefore wish to calculate the rotational and vibrational contributions to the internal energy,

$$U = U_{\text{trans}} + U_{\text{rot}} + U_{\text{vib}} + U_{\text{elec}} \qquad (16.66)$$

For a diatomic gas (rigid rotator) at around 100 K, the rotational partition function is given by Eq. (16.54):

$$u_{\text{rot}} = RT^2\left(\frac{\partial \ln Z_{\text{rot}}}{\partial T}\right)_V = RT^2\frac{\partial \ln T}{\partial T}$$

Hence,

$$\boxed{u_{\text{rot}} = RT} \qquad (16.67)$$

The vibrational partition function is given by Eq. (16.59); hence

$$u_{\text{vib}} = RT^2\left(\frac{\partial \ln Z_{\text{vib}}}{\partial T}\right)_V = RT^2\frac{\partial\left(-\dfrac{x}{2}\right)}{\partial T} + RT^2\frac{\partial \ln(1 - e^{-x})^{-1}}{\partial T}$$

where $x = hv/kT$, i.e.

$$u_{vib} = \frac{RT^2 hv}{2kT^2} + RT^2 \frac{e^{-x}}{1 - e^{-x}} \cdot \frac{hv}{kT^2}$$

Hence,

$$u_{vib} = \frac{Lhv}{2} + \frac{Lhv}{e^{hv/kT} - 1}$$

Alternatively, we have

$$u_{vib} = \frac{Lhv}{2} + \frac{RTx}{e^x - 1} \qquad (16.68)$$

or, relative to the vibrational ground level,

$$u_{vib} = \frac{RTx}{e^x - 1} \qquad (16.69)$$

Note that if hv is very large compared with kT, then the vibrational contribution to the molar internal energy reduces to

$$u_{vib} = \frac{Lhv}{2} \qquad (16.70)$$

This result can also be derived as follows. When the vibrational energy levels are measured relative to the ground level, i.e. putting $\varepsilon_0 = 0$, we know that the vibrational partition function is unity if $hv \gg kT$ (see Section 16.16). Making allowance for each of the L molecules in 1 mole possessing energy $hv/2$ which is unaccounted for, Eq. (16.70) is obtained for the total molar vibrational energy.

The kinetic theory of ideal gases shows that, when two molecules collide, the energy exchanged is of the order of kT. The energy required for a vibrational quantum jump is hv. Now if $hv \gg kT$, then it is improbable that a molecule will attain sufficient energy by collision to excite it vibrationally. If the temperature is sufficiently high, so that $hv = kT$, then the molecule may be excited to higher vibrational states. The temperature $\theta_V = hv/k$ is known as the *characteristic vibrational temperature*. For hydrogen, $\theta_V = 6340 \text{ K}$; for carbon monoxide, $\theta_V = 3120 \text{ K}$; and for chlorine, $\theta_V = 807 \text{ K}$. Similarly, for rotation, the energy required is about $h^2/8\pi I^2$. Hence, the *characteristic rotational temperature* is $\theta_R = h^2/8\pi I^2 k$ and below this temperature rotation is unlikely. For hydrogen, $\theta_R = 87 \cdot 2 \text{ K}$; for carbon monoxide, $\theta_R = 2 \cdot 76 \text{ K}$; and for chlorine, $\theta_R = 0 \cdot 62 \text{ K}$.

It can be shown mathematically that, as $x \to 0$, $x/(e^x - 1) \to 1$. Hence, from Eq. (16.69), the limiting vibrational contribution to the molar internal energy above the ground level is RT.

The internal energy of a diatomic molecule (all ground levels arbitrarily zero) may be found by substituting from Eqs (16.64), (16.67) and (16.68) into Eq. (16.66), giving

$$u = \tfrac{3}{2}RT + RT + \frac{RTx}{e^x - 1} \tag{16.71}$$

The rotational contribution to the molar heat capacity of an ideal diatomic gas is obtained by differentiating u_{rot} with respect to temperature at constant volume:

$$c_{V \, rot} = \left(\frac{\partial u_{rot}}{\partial T} \right)_V = R \tag{16.72}$$

The vibrational contribution to the molar heat capacity at constant volume is found by differentiating u_{vib} (Eq. 16.69), with respect to temperature at constant volume:

$$c_{V \, vib} = \left(\frac{\partial u_{vib}}{\partial T} \right)_V = \frac{Lh\nu e^{h\nu/kT}}{(e^{h\nu/kT} - 1)^2} \cdot \frac{h\nu}{kT^2} = \frac{Re^{h\nu/kT}}{(e^{h\nu/kT} - 1)^2} \left(\frac{h\nu}{kT} \right)^2$$

Hence,

$$c_{V \, vib} = \frac{Rx^2 e^x}{(e^x - 1)^2} \tag{16.73}$$

The molar entropy of a diatomic ideal gas is (Eq. 16.40)

$$s = R \ln \frac{Z}{L} + \frac{u}{T} + R$$

or,

$$s = R \ln \frac{Z_{trans} \cdot Z_{rot} \cdot Z_{vib} \cdot Z_{elec}}{L}$$

$$+ \frac{u_{trans} + u_{rot} + u_{vib} + u_{elec}}{T} + R$$

Rearranging,

$$s = R \ln \frac{Z_{trans} \cdot Z_{elec}}{L} + \frac{u_{trans} + u_{elec}}{T} + R$$

$$+ R \ln Z_{rot} + \frac{u_{rot}}{T} + R \ln Z_{vib} + \frac{u_{vib}}{T} \tag{16.74}$$

The first three terms have been calculated previously (see Section 16.18).

$$s_{rot} = R \ln Z_{rot} + \frac{u_{rot}}{T} = R \ln \frac{8\pi^2 I k T}{\sigma h^2} + R \qquad (16.75)$$

and $s_{vib} = R \ln Z_{vib} + \frac{u_{vib}}{T} = -\frac{Rx}{2} - R \ln (1 - e^{-x}) + \frac{Rx}{2} + \frac{Rx}{e^x - 1}$

or,

$$s_{vib} = -R \ln (1 - e^{-x}) + \frac{Rx}{e^x - 1} \qquad (16.76)$$

The other properties of the gas may now be determined.

16.20 Equilibrium constant of ideal-gas reactions

The van't Hoff isotherm states

$$-\Delta G^\ominus = RT \ln K_{P/P^\ominus}$$

Thus, if we can determine the standard Gibbs function change for a reaction, the equilibrium constant may be calculated. We need to calculate the molecular partition function in the standard state so that the standard Gibbs function may be obtained for each of the gases. From Eq. (16.43) we have

$$g^\ominus = -RT \ln \frac{Z^\ominus}{L}$$

The molecular partition function is

$$Z = Z_{trans} \cdot Z_{rot} \cdot Z_{vib} \cdot Z_{elec}$$

$$Z = Z_{rot} \cdot Z_{vib} \cdot Z_{elec} \cdot \frac{(2\pi m k T)^{\frac{3}{2}}}{h^3} v \qquad (16.77)$$

The standard state of an ideal gas is the state of the pure gas at a pressure of 101 325 N m^{-2} at the temperature under consideration. For an ideal gas, $v = RT/P$. Since $P^\ominus = 101\ 325$, $v = RT/101\ 325$ and we may write

$$Z^\ominus = Z_{rot} \cdot Z_{vib} \cdot Z_{elec} \cdot \frac{(2\pi m k T)^{\frac{3}{2}} RT}{101\ 325\ h^3}$$

Now consider a reaction A $\rightleftharpoons$ B. In order to consider equilibrium, we must calculate the molecular partition functions of A and B relative to the same energy zero. Figure 16.5 shows the energy levels of molecules A and B relative to the same zero energy.

However, it is more convenient to calculate the molecular partition function of each substance relative to its own ground level,

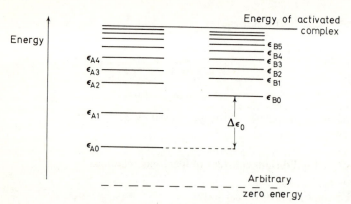

Figure 16.5. Energy levels for constituents A *and* B

setting the energy of each ground level to zero. Relative to the arbitrary zero, the lowest levels of molecules A and B are ε_{A0} and ε_{B0} and the corresponding partition function for molecule A is

$$Z'_A = \Sigma_i \, e^{-(\varepsilon_{Ai} + \varepsilon_{A0})/kT} = Z_A e^{-\varepsilon_{A0}/kT}$$

where Z'_A and Z_A are the molecular partition functions of A using the arbitrary zero and setting the ground level to zero, respectively. Hence, for the reaction under consideration, we have

$$\Delta G^\ominus = -RT \ln \frac{(Z_B^\ominus/L)}{(Z_A^\ominus/L)} \cdot e^{-\Delta \varepsilon_0/kT} \qquad (16.78)$$

where $\Delta \varepsilon_0 = \varepsilon_{B0} - \varepsilon_{A0}$. Comparing this with the van't Hoff isotherm,

$$K_{P/P^\ominus} = \frac{Z_B^\ominus}{Z_A^\ominus} \cdot e^{-\Delta \varepsilon_0/kT} \qquad (16.79)$$

Considering the more general reaction $aA + bB \rightleftharpoons cC + dD$, we obtain

$$K_{P/P^\ominus} = \frac{(Z_C^\ominus/L)^c (Z_D^\ominus/L)^d}{(Z_A^\ominus/L)^a (Z_B^\ominus/L)^b} \cdot e^{-\Delta \varepsilon_0/kT}$$

or,

$$K_{P/P\ominus} = \frac{(Z_C^\ominus)^c(Z_D^\ominus)^d}{(Z_A^\ominus)^a(Z_B^\ominus)^b} L^{-\Delta n} \cdot e^{-\Delta\varepsilon_0/kT} \qquad (16.80)$$

where $\Delta n = c + d - (a + b)$ and $\Delta\varepsilon_0$ is the difference in the ground level energies of the reactants and products.

16.21 Perfect crystals—the Einstein approximation

In a monatomic crystalline solid the atoms are arranged in a regular pattern and are fixed in position. Thus they may be regarded as being distinguishable and also as having no translational energy. Of course, the atoms are not independent of one another since strong internal forces exist. The statistical determination of the various properties is thus a many-body problem which cannot be solved. However, Einstein made the simplifying assumption that the small thermal vibrations of the atoms about their mean positions are independent and that the restoring forces obey Hooke's law—that is, the restoring force is proportional to the displacement from the mean position, and therefore the particles behave as harmonic oscillators.

To specify the instantaneous position of an atom relative to its mean position, three coordinates, x_1, y_1, z_1, are required. Hence, for a crystal with N identical atoms, $3N$ coordinates are required. The vibration of each particle may be considered in terms of vibrations along the three coordinate axes. Therefore the crystal may be considered to consist of $3N$ independent harmonic oscillators.

Einstein further assumed that the $3N$ oscillators all had the same frequency, v.

From quantum mechanics it is known that the allowable energy levels of a harmonic oscillator of frequency v are given by

$$\varepsilon_n = (n + \tfrac{1}{2})hv$$

where $n = 0, 1, 2, \ldots$

No degeneracy exists unless different orientations of the oscillators are possible.

If the energy is measured relative to the ground level ($n = 0$), we have

$$\varepsilon'_n = nhv \qquad (16.81)$$

As previously mentioned, the vibration of an atom may be considered in terms of three vibrations at right angles to one another.

The permissible energies for these harmonic oscillations must be given by Eq. (16.81) and the partition function by Eq. (16.60) – that is

$$Z_{vib} = (1 - e^{-x})^{-1}$$

where $x = hv/kT$.

The atomic partition function is

$$Z = Z_{vib} \cdot Z_{vib} \cdot Z_{vib} = (1 - e^{-x})^{-3}$$

The molar internal energy relative to the ground level (see (Eq. 16.36) is

$$u = RT^2 \left(\frac{\partial \ln Z}{\partial T} \right)_V$$

Hence,

$$u = \frac{3RTx}{e^x - 1} \tag{16.82}$$

The molar heat capacity at constant volume (see Eq. 16.38) is found to be

$$c_V = \frac{3Rx^2 e^x}{(e^x - 1)^2} \tag{16.83}$$

As T becomes larger, so that $kT \gg hv$ and $x \to 0$, c_V tends to $3R$. This is in good agreement with Dulong and Petit's rule, according to which solids have a specific heat of 25 J K^{-1} mol^{-1}. (Remember $e^x = 1 + x/1! + x^2/2! + x^3/3! + \ldots$ and $x(e^x - 1)^{-1} \to 1$ as $x \to 0$). As T approaches 0 K, c_V tends to zero. Although good agreement is obtained with experimental data, the theoretical values of c_V at low temperatures are too small. Of course, in truth, the oscillators are not harmonic and independent, nor do they all have the same frequency.

16.22   The Debye approximation

By considering a solid to be a continuous elastic medium, Debye derived an expression for the distribution of the frequencies of the oscillators. The theory showed that a broad spectrum of frequencies existed which finished abruptly at a maximum value v_{max}. Debye showed that

$$u = 9RT \left(\frac{T}{\theta_D}\right)^3 \int_0^{x_{max}} \frac{x^3}{e^x - 1} \, dx \qquad (16.84)$$

where $x = hv/kT$, $x_{max} = hv_{max}/kT$ and $\theta_D = hv_{max}/k$. θ_D is known as

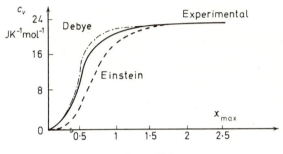

Figure 16.6.

the Debye temperature of a solid. Differentiating Eq. (16.84) with respect to temperature at constant volume,

$$c_V = 9R \left(\frac{T}{\theta_D}\right)^3 \int_0^{x_{max}} \frac{x^4 e^x}{(e^x - 1)^2} \, dx$$

Integrating by parts,

$$c_V = -\frac{9Rx_{max}}{e^{x_{max}} - 1} + \frac{36R}{x_{max}^3} \int_0^{x_{max}} \frac{x^3}{e^x - 1} \, dx \qquad (16.85)$$

At high temperatures, $x_{max} \, (= \theta_D/T)$ is small, and it can be shown that the value of c_V equals the classical value $3R$. At low temperatures, x_{max} is large and the first term of Eq. (16.85) is negligible. Also, with only very small error, the upper limit of the integration may be replaced by infinity, the value of the integral then being $\pi^4/15$. Hence, for low temperatures,

$$c_V = \frac{36\pi^4 R}{15 x_{max}^3} = \frac{12\pi^4 R}{5} \left(\frac{T}{\theta_D}\right)^3 \qquad (16.86)$$

This is known as the Debye T^3 law. Comparison between the Debye and Einstein approximations is given in Figure 16.6.

16.23   The partition function of the system

So far we have considered a single molecule and derived the properties of the system from the molecular partition function, Z, for the state of the system. An alternative approach which is often used is to consider the whole system and the energy levels available to it (including the corresponding degeneracies) in order to obtain the *partition function of the system*, Q. Maxwell–Boltzmann statistics are generally used to derive the properties of the system.

The energy of the system, E_i, is equal to the sum of the energies ε_i of the non-interacting particles

$$E_i = \varepsilon_1(1) + \varepsilon_2(2) + \varepsilon_3(3) + \ldots + \varepsilon_N(N)$$

That is, particle 1 has energy ε_1, particle 2 has energy ε_2, etc. Since each particle is identical, they all have the same allowable energy levels. Now each different way of assigning the particles among these energy levels results in a new allowable energy level of the whole system. Hence, the partition function of the system

$$Q = \Sigma_i e^{-E_i/kT}$$

may be written as

$$Q = \Sigma_i e^{-[\varepsilon_i(1) + \varepsilon_i(2) + \ldots + \varepsilon_i(N)]/kT}$$

or,

$$Q = \Sigma_i e^{-\varepsilon_i(1)/kT} \cdot e^{-\varepsilon_i(2)/kT} \ldots e^{-\varepsilon_i(N)/kT}$$

where $\varepsilon_i(1)$ is the ith energy level of particle 1 and the degeneracies are omitted for convenience. Now, since all the particles have the same energy levels, we have

$$Q = (\Sigma_i e^{-\varepsilon_i/kT})^N = Z^N \tag{16.87}$$

This applies to the case where particles are distinguishable, as in crystals. If the particles are indistinguishable, as in a gas, then the partition function of the system is

$$\boxed{Q = \frac{Z^N}{N!}} \tag{16.88}$$

since the number of permutations of N particles among themselves is $N!$.

Equation (16.87) could have been derived directly by considering the partition function of the system to be the product of all the N molecular partition functions:

$$Q = Z_1 . Z_2 . Z_3 ... Z_N$$

Since all the particles are identical, Eq. (16.88) is obtained.

The various molar properties in terms of Q are

$$u = kT^2 \left(\frac{\partial \ln Q}{\partial T} \right)_V$$

$$c_V = \frac{k}{T^2} \left[\frac{\partial^2 \ln Q}{\partial \left(\frac{1}{T} \right)^2} \right]_V$$

$$s = k \ln Q + kT \left(\frac{\partial \ln Q}{\partial T} \right)_V$$

$$a = -kT \ln Q$$

and

$$g = -kT \ln Q + RT$$

Substituting for Q from Eq. (16.88) and using Stirling's formula in these equations, the corresponding equations obtained previously in Section 16.11 are derived.

Problems

$$R = 8 \cdot 314 \, \text{J K}^{-1} \, \text{mol}^{-1} \qquad k = 1 \cdot 381 \times 10^{-23} \, \text{J K}^{-1}$$
$$h = 6 \cdot 626 \times 10^{-34} \, \text{J s} \qquad L = 6 \cdot 0225 \times 10^{23} \, \text{mol}^{-1}$$

1. Calculate the entropy of argon gas at its boiling point, 87·3 K, given that the mass of an argon atom is $6 \cdot 63 \times 10^{-26}$ kg and the molar volume of argon gas at 87·3 K is $7 \cdot 159 \times 10^{-3} \, \text{m}^3$.
Answer 129 J K^{-1} mol^{-1}

2. Calculate the rotational contribution to the entropy of carbon dioxide at 298 K, given that the moment of inertia is $1 \cdot 449 \times 10^{-46}$ kg m^2.
Answer 57·23 J K^{-1} mol^{-1}

3.   Bearing in mind the answer to Problem 2 above, show that the vibrational contribution to the entropy of carbon dioxide at 298 K is negligible if the vibrational energy levels are given by $\varepsilon = 4 \cdot 30(v + \frac{1}{2}) \times 10^{-20}$ J.

4.   Calculate the entropy of nitrogen (which is a symmetrical molecule) at 298 K, given that the ground level is a singlet. The molecular weight is 28·02, $I = 1 \cdot 381 \times 10^{-46}$ kg m^2, $x = 11 \cdot 39$.
Answer   191·5 J K^{-1} mol^{-1}

5.   Calculate the entropy of hydrogen chloride at 298 K, given that the ground level is a singlet, the molecular weight is 36·47, and $I = 2 \cdot 65 \times 10^{-47}$ kg m^2, the vibrational contribution being negligible.
Answer   187·3 J K^{-1} mol^{-1}

Appendix 1
Units

Special notes on the use of SI units

Only the singular form of units is to be used, e.g. km *not* kms

Full stops at the end of abbreviations are to be omitted, e.g. km *not* km.

Digits should be grouped in threes about the decimal point in order to facilitate the reading of long numbers. Commas should not be used to space digits in numbers, e.g. 16 543 211·133 45 *not* 16,543,211·133,45

The degree sign is to be omitted when the Kelvin scale is employed, e.g. 273 K *not* 273°K

Where used, no more than one solidus should be employed, e.g. J/kg K *not* J/kg/K

Basic units

The following are the basic units of SI:

Physical quantity	Unit	Symbol
length	metre	m
mass	kilogramme	kg
time	second	s
electrical current	ampere	A
temperature	kelvin	K
luminous intensity	candela	cd
plane angle	radian	rad
solid angle	steradian	sr

The mole has been recommended as a basic unit but remains to be accepted.

Derived units

Derived units with special names are as follows:

Physical quantity	Name	Symbol
energy	joule	J
force	newton	N
power	watt	W
electric charge	coulomb	C
electrical resistance	ohm	Ω
electrical potential difference	volt	V
electric capacitance	farad	F
magnetic flux	weber	Wb
magnetic flux density	tesla	T
inductance	henry	H
luminous flux	lumen	lm
illumination	lux	lx
frequency	hertz	Hz

Prefixes

Prefix	Symbol	Factor
tera	T	10^{12}
giga	G	10^{9}
mega	M	10^{6}
kilo	k	10^{3}
†hecto	h	10^{2}
†deca	da	10^{1}
†deci	d	10^{-1}
†centi	c	10^{-2}
milli	m	10^{-3}
micro	μ	10^{-6}
nano	n	10^{-9}
pico	p	10^{-12}
femto	f	10^{-15}
alto	a	10^{-18}

†The use of these prefixes is not recommended and is to be limited to occasions when other prefixes are inconvenient.
  Only one prefix should be used for a given unit.

EXAMPLES

$$1\ 000\ 000\ \text{m} = 1\ \text{Mm}\ not\ 1\ \text{kkm}$$
$$1000\ \text{kg} = 1\ \text{Mg}\ not\ 1\ \text{kkg}$$
$$10^{-9}\ \text{kg} = 1\ \mu\text{g}\ not\ 1\ \text{nkg}$$

The power to which a unit is raised applies to the whole unit including the prefix.

EXAMPLES

$$\text{km}^2 = (\text{km})^2 = (1000\ \text{m})^2 = 10^6\ \text{m}^2\ not\ 1000\ \text{m}^2$$
$$\mu\text{m}^3 = (\mu\text{m})^3 = (10^{-6}\ \text{m})^3 = 10^{-18}\ \text{m}^3\ not\ 10^{-6}\ \text{m}^3$$

Conversions

Physical quantity	Unit	SI equivalent
length	†ångstrom	10^{-10} m
	†inch	0·0254 m
	mile	1·609 34 km
area	†barn	10^{-28} m^2
	†square inch	645·16 mm^2
volume	cubic inch	$1·638\ 71 \times 10^{-5}$ m^3
	†litre	10^{-3} m^3
mass	†pound	0·453 592 37 kg
force	†dyne	10^{-5} N
	poundal	0·138 255 N
	pound-force	4·448 22 N
	†kilogramme force	9·806 65 N
pressure	†atmosphere	101·325 kN m^{-2}
	†bar	10^5 Nm^{-2}
	torr	133·322 N m^{-2}
	p.s.i.	6894·76 N m^{-2}
energy	†calorie, international	4·1868 J
	†calorie, 15°C	4·1855 J
	†calorie, thermochemical	4·184 J
	B.t.u.	1055·06 J
	electron volt	$1·6021 \times 10^{-19}$ J
	†erg	10^{-7} J
power	horse power	745·700 W

†These conversions are exact.

Physical constants	*Value*
Gas constant	$8\cdot3143\ \mathrm{J\ K^{-1}\ mol^{-1}}$
Molar volume of ideal gas at 101 325	
N m^{-2}(= 1 atm) and 273·15 K	$2\cdot241\ 36 \times 10^{-2}\ \mathrm{m^3\ mol^{-1}}$
Avogadro constant	$6\cdot022\ 52 \times 10^{23}\ \mathrm{mol^{-1}}$
Faraday constant	$9\cdot648\ 70 \times 10^{4}\ \mathrm{C\ mol^{-1}}$
Planck constant	$6\cdot6256 \times 10^{-34}\ \mathrm{J\ s}$
velocity of light	$2\cdot997\ 925 \times 10^{8}\ \mathrm{m\ s^{-1}}$

Further reading

Anderton, P. and Bigg, P.H., *Changing to the Metric System*, H.M.S.O. London (1969)

Socrates, G. and Sapper, L. J., *SI and Metrication Conversion Tables*, Newnes–Butterworth, London (1969)

Appendix 2
Symbols

A	Helmholtz function
a	activity
$a_\pm$	mean ionic activity
C_P	heat capacity at constant pressure
c_P	molar heat capacity at constant pressure
C_V	heat capacity at constant volume
c_V	molar heat capacity at constant volume
E	e.m.f.
f	fugacity
f_i	fugacity of component i
G	Gibbs function
g	molar Gibbs function
H	enthalpy
h	molar enthalpy
$K_{P/P^\ominus}$	equilibrium constant for reaction in perfect gaseous mixture
L	Avogadro's constant
M	mass
n	amount of substance in moles
P	pressure
q	heat absorbed
R	universal gas constant
S	entropy
s	molar entropy
T	temperature
U	internal energy
u	molar internal energy
V	volume
v	molar volume
w	work done
x	mole fraction
Z	molecular partition function
α	isothermal expansivity

β	isothermal bulk modulus
ε_i	ith energy level
γ	activity coefficient
$\gamma_\pm$	mean ionic activity coefficient
η	efficiency
κ	isothermal compressibility
μ_i	chemical potential of substance i
μ_{JK}	Joule–Kelvin coefficient
Π	product
Σ	sum
ϕ	fugacity coefficient
ϕ_i	fugacity coefficient of component i
$\ominus$	standard state
$\oint$	integral over the cycle
$\mathrm{d}x$	exact differential of x
$\text{đ}y$	inexact differential of y

Lower case letters are used for molar or specific quantities, e.g. H enthalpy, h molar enthalpy.

Index